PUDDLE QUESTIONS

GRADE **6**

Assessing Mathematical Thinking

Joan Westley

 Creative Publications

Acknowledgments

Writers: Susan Greer, Nancy Tune
Editors: Sarah Le Forge, Mary Scott Martinson, and Nancy Tune
Classroom Coordinators: Margaret Hall, Nancy Mack, Cindy Reak, Julie Ruskey, Lyn Ungrodt, Avery Walker, Laurie Waring, and Linda Welsh
Spanish Translation: OSO Publishing Services
Creative Director: Ken Shue
Production Coordinator: Joe Shines
Design and Production: London Road Design
Cover Art and Illustrations: Jim M'Guinness
Photographs: Steve McCue, Michael Carr

With special thanks to Valerie Stofac, Heather McDonald, and Cindy Reak for their advice, support, and confidence in the value of this project.

Seventy-six teachers in twenty-one schools in four states field tested the problems in this series with approximately twenty-five hundred students. Heartfelt thanks to every one of you—especially the students, whose responses we treasure.

Participating Schools:

California
Barnard-White, Union City
Brookvale, Fremont
Keys, Palo Alto
Kynoch, Marysville
Maloney, Fremont
Millard, Fremont
Ohlone, Palo Alto
Paradise, Paradise
Pines, Magalia
Ponderosa, Paradise
Walters, Fremont
Warm Springs, Fremont

Florida
Eastbrook, Winter Park
English Estates, Fern Park
Greenwood Lakes, Lake Mary
Hamilton, Sanford
Red Bug, Casselberry
Rock Lake, Longwood
Sterling Park, Casselberry

Illinois
Hazel Dell, Springfield

Wisconsin
Wingra, Madison

Photographs:
Monroe Middle School, San Jose, California
Warm Springs School, Fremont, California

© 1994 Creative Publications
1300 Villa Street
Mountain View, California 94041–1197

All rights reserved. No part of this publication may be reproduced, stored in a retrieval system, or transmitted, in any form or by any means, electronic, mechanical, photocopying, recording, or otherwise, without the prior written permission of the publisher. Printed in the United States of America.

Puddle Questions is a registered trademark.

ISBN: 1–56107–345–8

6 7 8 9 10 9 8

Table of Contents

What Is *Puddle Questions*™? ... 4
Presenting the Problems ... 5
Assessing the Work ... 7
Extending the Learning ... 8
Generalized Scoring Rubric ... 9

Investigation 1: The Puddle Problem
Focus: Measurement ... 10

Investigation 2: Television Survey
Focus: Statistics ... 20

Investigation 3: One to a Million
Focus: Estimation and Reasoning ... 28

Investigation 4: Equations Galore!
Focus: Arithmetic Operations ... 36

Investigation 5: Doghouse Designs
Focus: Design and Construction ... 44

Investigation 6: Telephone Talk
Focus: Geometry and Language ... 52

Investigation 7: Is It Fair?
Focus: Probability ... 60

Investigation 8: Lemonade Stand
Focus: Situational Problem Solving ... 68

Bibliography ... 76
Translations of Student Work ... 76
Student Pages Blackline Masters ... 77
Writing Math Reports ... 93
Self-Assessment Key ... 94
Observation Sheets ... 95
Record Sheet ... 96

What Is Puddle Questions™?

In 1992, when I was working with a K–2 classroom on some theme-based activities, we asked the students this question: "How would you go about measuring a puddle?" We had the students write their ideas on paper, using pictures and words. What came from those students that day was so interesting that it gave birth to this series—Puddle Questions: Assessing Mathematical Thinking.

When we looked at the students' responses, we discovered that their answers told us more about what they knew about measurement than any test we could have devised. Their responses were revealing in a variety of ways. They gave us information about how individual students were able to think through a complex problem, plan a solution, use tools and techniques, and communicate their thinking so others could understand it.

We decided to put together a series of open-ended, thought-provoking problems similar to the Puddle Problem for each grade level, 1 through 6. The hardest part was finding the right questions. We wanted problems that would give us a chance to assess students' thinking about broad mathematical ideas such as measurement or probability. The problems needed to pose authentic questions about mathematics in the real world. To be Puddle Questions, the problems needed to be open-ended, with no one right or best answer. Like real-life problems, Puddle Questions needed to be "messy" in the sense that not all aspects of the problem were clearly defined. But they also needed to be scorable, so that we could use the responses to assess students' mathematical thinking.

Once we had developed the problems, we piloted them in classrooms across the country. Over 70 classrooms participated. When the packages of student responses came back to us, we delighted in the variety of unique ways students found to solve the problems and present their thinking on paper. We are proud to share their work with you in this book.

As we worked with students and teachers on Puddle Questions, we found that the initial phase was often difficult. Students were not used to being asked to solve these kinds of problems. After years of multiple-choice tests and fill-in-the-blank worksheets, students were often unsure of themselves when tackling bigger questions. But after a few investigations, the students seemed to thrive on the problems, and produced work that far exceeded our expectations.

I hope you get as much joy as I have from seeing students grow in their ability to solve rich and complex problems.

Sincerely,

Joan Westley

Presenting the Problems

The Teacher's Role
With *Puddle Questions,* as with open-ended assessment in general, the teacher acts as a facilitator in the problem-solving process. Teachers should hand out the problem pages and introduce the problem briefly by reading it aloud. (Blackline masters of the problem pages in both English and Spanish are provided on pages 77–92.) Students need to understand that you want to see their best work and their most careful thinking. They should know that there is no one right answer to the problem. They should also know what they will be evaluated on in the assessment task. For each problem, assessment criteria are listed in the Presenting the Problem section of the teaching notes. You can also ask students the questions found in Prompts for Getting Started.

While the students work on the problems, you can reread the Prompts for Getting Started questions, either to individuals or to the whole group. This will help remind students of aspects of the problem they may not have attended to at first, but that may be more pertinent as they get deeper into the problem.

Your most important role while the students are working is to sit back and let them work through the problem themselves. Try not to direct their work into one avenue or another. Give the students time to come up with their own ideas.

Difficulty Level
Let the students know that these investigations are difficult. They're intended to expose students to challenging ideas so you can see how they respond. All students should be able to respond in some way. Because they are allowed to respond to a question in any way they see fit, you get a good look at what concepts students have internalized.

Time Period
Most problems take one class period on average, but students should be allowed to work on the problems as long as they want to. They may wish to work for a while and then revise their papers or start again.

Materials
Choosing the proper tools is part of the solution process. Students should have access to any reasonable tools or equipment they need to solve the problem. Calculators, manipulatives such as linking cubes or Base Ten Blocks, and measuring devices such as rulers or scales should be readily available as appropriate. The important thing is to avoid dictating what tools the students use. For example, distributing graph paper for a problem might give the false impression that graph paper must be used for a correct solution. A better method is to have available a range of standard equipment that students can choose from. A list of tools and materials is provided in the introduction to each problem.

Grouping Arrangements

A suggested grouping arrangement is given for each problem. If students work in pairs or small groups, each student is still responsible for writing up and explaining his or her own solution.

The Students' Role

In most problems, students are asked to write a report to explain their thinking and reasoning, to draw pictures or sketches of their plans, or to display data in some way. Let the students use their own invented spellings for words they do not know. Writing in a primary language other than English is also a good idea, if the student would be handicapped writing in English. If work is difficult to decipher, you may want the student to read the paper to you so you can understand what is being said. Students can also dictate their ideas for you or an aide to write on their papers, if the technical difficulties of writing are likely to inhibit their responses. This is often the case at first grade, and sometimes at second grade.

In *Puddle Questions,* we are asking students to take responsibility for doing their best work and for showing what they know. This mirrors life in the workplace. There, your tasks are often not spelled out for you. Rather, you are expected to think through a problem carefully and present your ideas thoughtfully. You are expected to communicate in a way that gives people a clear understanding of your thinking. Often, you are expected to display information in interesting and easily understandable ways.

You can help students keep these things in mind by giving them a copy of page 93, Writing Math Reports. This gives students some guidelines as they write up their reports for the investigations. The page could be stapled to their math portfolios and referred to each time they do an investigation. It could also be used as an overhead transparency and displayed while students work on a problem.

Students who finish early should be asked to review the question and their answer to make sure they have done as much with the problem as they can. If they are truly finished, direct them to do something quiet while others finish. When everyone has finished, students can clip or staple their problem page to their report and hand the pages in together.

Self-Assessment

On the students' problem page is a self-assessment dial. Students are to shade the dial to show the level at which they feel they performed in the investigation. A Self-Assessment Key provided on page 94 of this book explains how to fill in the dial to show low, medium, high, and exceptional responses. It's interesting to compare students' self-assessment scores to the scores they receive from you for the investigation. You might want to discuss discrepancies between the two should they occur. The Self-Assessment Key can be used as an overhead transparency, or you can give each student a sheet to staple onto the investigation portfolio.

Assessing the Work

Scoring the Responses

We recommend a holistic approach for scoring the responses to the investigations. In a holistic approach, you look at the work globally and give one overall score to the response, rather than giving points for various aspects of the response. The emphasis should be on thinking skills, mathematical understanding, and reasoning rather than on use of arithmetic or computational techniques. Thus, a student who has made arithmetic errors might receive a high score because of the depth of mathematical thinking that the response shows. Conversely, a student with accurate computation may receive a low score because the response does not show an understanding of mathematical ideas. Exceptional scores are reserved for the one or two responses that exceed expectations. You may find that no student's response fits in this category for a given investigation.

You can score your students' work using one of the following methods:

- The scoring rubric provided with the problem. This rubric provides four levels of response to the specific problem in the investigation: Low, Medium, High, and Exceptional. You may want to add N/A (no attempt) and INC (incomplete) categories to the rubric.

- The Generalized Scoring Rubric on page 9. This rubric, which can be applied to all the problems in the series, has the same four levels of response: Low, Medium, High, and Exceptional.

- The four-piles method. In this method, the scorer reads the responses and places them in one of two piles: high or low. Then the scorer further divides each pile into two piles, forming four levels of response to the problem.

- Your district's own scoring rubric or a scoring rubric of your choice. Many districts or states have established their own assessment schemes and scoring rubrics involving anywhere from three to six levels. In general, to create a three-point rubric from the *Puddle Questions* rubric, eliminate the Exceptional category. To create a five-point rubric, divide the Medium category in two. To create a six-point rubric, divide both the Medium and the Low categories in two.

Student Responses

Shown with each problem are sample responses for each level: Exceptional, High, Medium, and Low. Some responses were gathered at the beginning of the school year, some at the middle, and some at the end. No doubt the same students would have done different work at different times. It is not possible to represent here the full range of responses for these types of problems. However, studying these samples will give you an idea of the variety of ways students can solve the problems. The samples may also help you to score your class's responses by providing one or two benchmarks for each level. Don't be alarmed, however, if your class's responses do not match the samples. Every classroom's responses are unique. How you score your students' work should be an individual decision, based on their needs and the amount of experience they have had with open-ended investigations.

A class record-keeping form is provided on page 96 of this book.

Making Observations
The scoring rubric gives only one part of the story of what you learn from a student's work on the investigations. To get a more complete picture, you'll want to make anecdotal records of your observations about how students tackled the problems, and what you learned about what they know. This can be done on plain note paper, 3" x 5" cards, or on the special Observation Sheets provided on page 95. Your observations should focus on these aspects of the students' responses: communication (how they explain their thinking), attitude (how they view the problem and its solution), collaboration (how they work with others on the problem), use of tools and equipment, math understanding, and reasoning. You might want to focus your observations on one of these areas each time you do an investigation.

Portfolios
Students can keep their investigation reports in a special portfolio. This way they can see improvement over the year in their ability to tackle open-ended problems. These portfolios are also useful for parent conferences. Authentic assessment is often very useful in showing parents what students can and cannot do.

Extending the Learning

Class Discussion and Sharing
The investigations are intended to be more than just an evaluation of what students can do. They are supposed to be a learning experience. To enhance the learning potential of each investigation, you'll want to follow up with a chance for students to discuss and share the work they did and the variety of ways they solved the problem. Suggestions are given for class discussions for each problem. The key is to let students explain their own thinking to each other. In doing so, they may (1) realize that they were misdirected in their reasoning, or (2) help others understand how they might have done the problem differently.

The Follow-Up Activity
The follow-up activity described for each problem is a natural outgrowth of the investigation itself. Its purpose is again to extend the learning experiences inherent in the problem situation. These activities usually take one class period to complete.

GENERALIZED SCORING RUBRIC

Responses in each category show some of the following characteristics:

Low Response
- Shows little grasp of the concepts.
- Fails to address significant aspects of the problem.
- Has major errors.
- Communicates poorly.
- Does not explain thinking.

Medium Response
- Shows some understanding of the concepts.
- Contains a complete response.
- Communicates unclearly or inappropriately.
- Explains thinking adequately, with little detail.

High Response
- Shows understanding of the concepts.
- Contains a complete response.
- Shows creativity.
- Communicates effectively.
- Explains thinking clearly.

Exceptional Response
- Contains all the characteristics of a high response.
- Goes beyond the requirements of the problem.
- Shows original thought.
- Gives strong supporting arguments.
- Explains thinking coherently and unambiguously.
- Shows exceptional mathematical thinking for the grade level.

The Puddle Problem

The Puddle Problem

Focus: Measurement

This investigation focuses on students' ideas about measurement. Your students have probably never been asked to think about measuring a puddle before. This is not a straightforward task like finding the radius of a circle pictured in a textbook; students have to decide for themselves what dimensions of a puddle to measure and how to measure them. It is interesting to see the varied and creative ways they come up with to measure a puddle—such as scooping the water out with a vacuum hose, or finding out how long it takes to walk around it.

Mathematical Ideas

- Measuring length, width, depth, perimeter, area, capacity
- Using measurement tools

Tools and Equipment

◆ blank paper
◆ lined writing paper
◆ pencils, crayons, markers

Grouping Arrangement

Individuals or small groups

Blackline Master in English, page 77 and in Spanish, page 78

Presenting the Problem

1. Pose this question to the students: **Suppose you wanted to tell someone how large a puddle is. What are some different ways you could do it?**

2. Tell the students to write a report on all the different ways they thought of to measure the puddle. **Your report should show the measuring tools you would use and tell how you would go about taking the measurements. You may need to explain how you would calculate the measurements. Be sure to use sketches wherever you think they would help.**

3. Tell students that the next time it rains, they will have a chance to measure puddles in the ways they have described.

Assessment Criteria

It is important to let students know what they will be evaluated on in this assessment. Tell them that you are interested in finding out

- ✔ how many different dimensions they think of measuring
- ✔ what tools they would use for measuring
- ✔ how clearly they describe their methods of measuring

Prompts for Getting Started

Students who have difficulty knowing how to approach this task could be prompted with questions like these:

- **What is one way you might measure a puddle?**
- **Can you think of any other ways?**
- **What will you use to do your measuring?**
- **How can you show all the different ways? Would making sketches help?**

If this is your students' first experience with an open-ended assessment question, you may need to give them extra encouragement. They may not understand the task or feel uncertain about what they are supposed to do. It may take a few investigations before they feel comfortable with the vagueness of these kinds of problems, especially with the notion of being responsible for communicating their thinking in writing. Once they have tackled a few of the investigations, their written reports will improve considerably.

The Puddle Problem

Assessing the Work

In this assessment, we are interested in finding out what students know about measurement. What tools do they use—rulers? measuring cups? scales? What dimensions do they measure—length? depth? diameter or radius? circumference? volume? area? What they report on is usually what they have internalized from all the work they have done in measurement.

Questions to ask yourself while scoring a response:

- How many different dimensions does the student plan to measure?
- What measurement tools does the student use? Are the tools appropriate for the measuring task?
- How clearly does the student communicate a plan for measuring the puddle?

SCORING RUBRIC

Low Response
Measurement methods proposed are unclear or unworkable. Explanation is missing or limited, and the response may be difficult to understand. Some basic misconceptions about measurement may be revealed.

Medium Response
Measurement methods are fairly clearly presented; they tend to be the most obvious choices. Explanations are adequate, with limited details.

High Response
The measurements and tools are clearly presented and may include some original ideas. The response shows a good understanding of the problem. Explanations are clear, with some details.

Exceptional Response
The methods for measuring go beyond the most obvious to include strategies that show original thought for the student's age. The response reveals an ability to think through a complex problem. Explanations are clear and effective, and include relevant details.

Measurement

1. If you wanted to mesur the legnth of the puddle you could take a tape mesure and take the legnth from end to end.
2. If you wanted to mesur the with of the puddle you would take a tape mesur and mesure it from side to side.
3. To mesur the amount of liquid in the puddle you would siphon the water into a gradguated cilender and see how many ML. Their was.
4. To mesur the depth of the water you could take a ruler and see how many cementers deep it is.
5. If you wanted to know the weight of the puddle you would siphon the water into a cup weigh it then pour out the water and weigh the cup then minus the weight of the cup from the water.
6. If you wanted to see the density you would find the mass then the volume then divide the mass by volume.
7. If you wanted to find the amount of dirt in the puddle you would boil the puddle until it all evaporated then weight the amount of remaning dirt.
8. If you wanted to mesur the solubility of salt in the puddle you would siphon the water into a beaker then stir in salt until it no longer disolved but first weigh the amount of salt you had when you started the experment and weigh it when you were done and minus the amounts and the remaning number would be the number of grams of salt in that could disolve in that puddle

Exceptional Response

This student goes beyond length, width, depth, weight, and capacity to describe how he would calculate the density and solubility of the puddle, bridging science and mathematics in his descriptions.

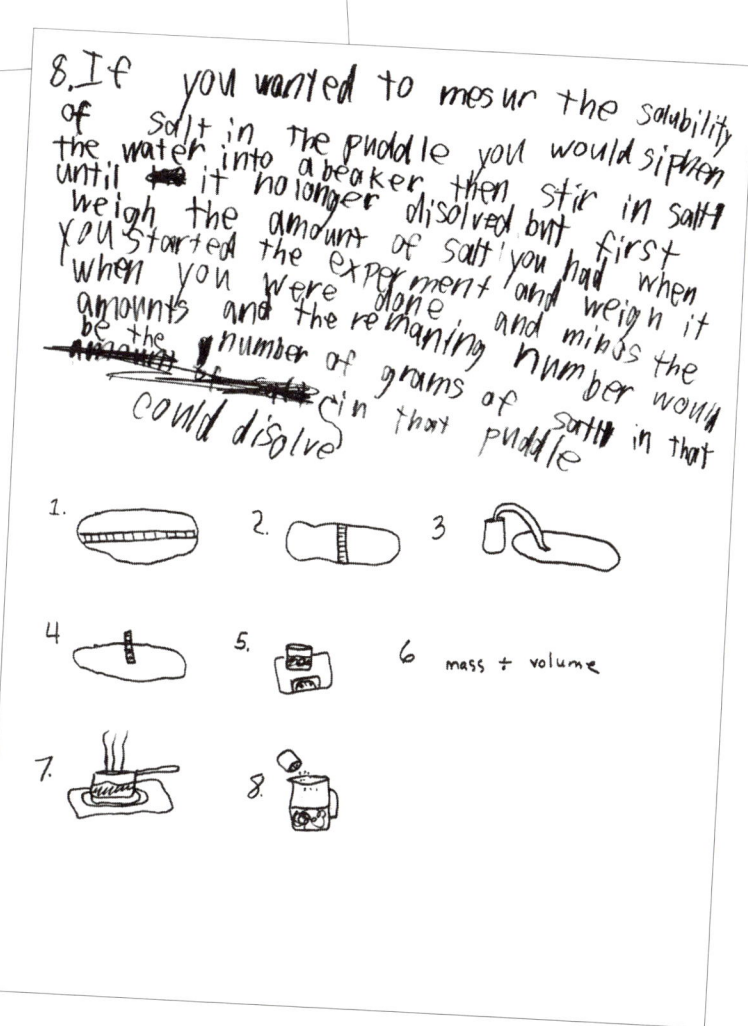

13

The Puddle Problem

High Response

This student's four ideas are clearly presented in an organized format. Her fourth idea, measuring the rate of evaporation, is especially interesting.

Measuring a Puddle

1) To measure the perimeter:

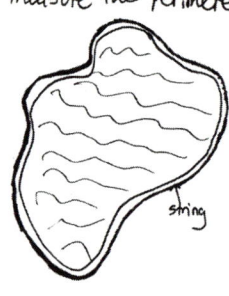

You would measure the perimeter of a puddle by putting a string around the edge of the puddle and then measuring the length of the string.

2) To measure the depth:

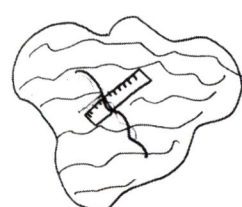

You would measure the depth of a puddle by sticking a ruler into the deepest part of the puddle, and then where the wet mark is, that is the depth of a puddle.

Measuring a Puddle

3) To measure the temperature of the puddle:

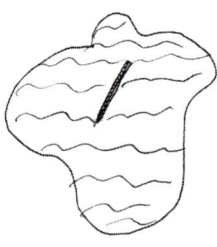

You would measure the temperature of the puddle by sticking a thermometer into the puddle and seeing what it is after a minute or so.

4) To measure how much the puddle will evaporate in one day:

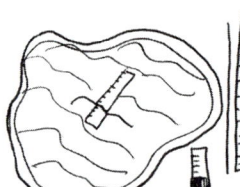 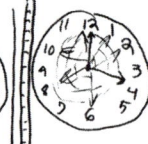  

To measure how much the puddle will evaporate in one day, you would measure the depth and perimeter (see #1 + #2), wait one day, measure the depth and perimeter again, and see how much both decreased.

Measurement

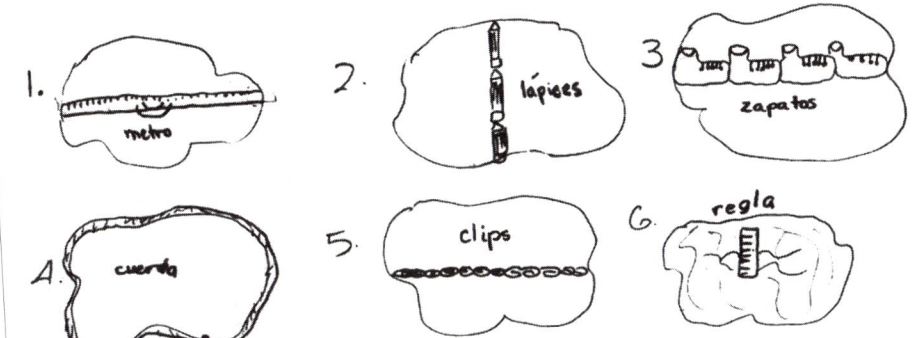

1. metro - yo usaría un metro para medir a través del charco.
2. lápices - yo usaría lápices para medir el ancho del charco.
3. zapatos - yo usaría zapatos para medir a través del charco.
4. cuerda - yo usaría una cuerda para medir alrededor del charco.
5. clips - yo usaría clips para medir através del charco.
6. regla - yo usaría una regla para ver qué tan hondo es el charco.

English translation, page 76

Medium Response

In this response, six ideas for measuring a puddle are all presented fairly clearly, with the help of some wonderful sketches. The student uses standard and nonstandard techniques.

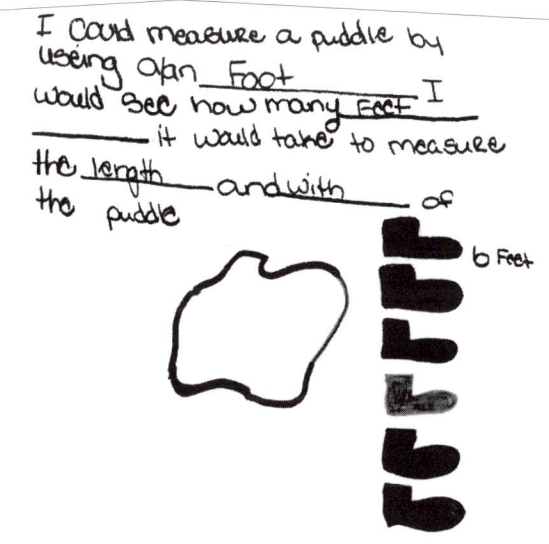

I could measure a puddle by useing a/an Foot I would see how many Feet it would take to measure the length and with of the puddle

6 Feet

Low Response

This student is so unused to responding to an open-ended investigation that she has turned her answer into a fill-in-the-blank worksheet. She has only two ideas for measuring a puddle. And her color-coded art, which indicates she is measuring circumference, is at odds with her writing.

I could measure a puddle by useing a Ruler I would see how many Inches it would take to measure the length and with of the puddle

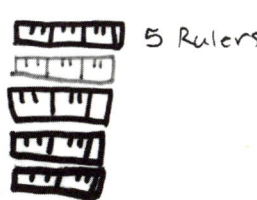

5 Rulers

15

The Puddle Problem Across the Grades

Giving the same question to students from kindergarten on up through the grades is a powerful way to see what our students are learning, and how they are growing in their ability to show what they know. The puddle question—*How would you measure a puddle?*—is interesting and challenging at all grade levels.

Grades K–1

At this level, students depend greatly on their pictures to tell what they are thinking. Responses may be dictated as this one is, or the students may use invented spelling. Their measuring ideas often include the kinds of nonstandard units they have been using to measure things in their classrooms.

Grades 2–3

The responses in the late primary grades are more detailed in their descriptions. The students are beginning to think about how they can communicate better so the reader will understand what they are thinking.

Measurement

1. Tape Measure — If you round off the puddle and measure it with a tape measure it would be really easy!! Say this was the puddle before it was rounded off.

This is the puddle after it's rounded off

Now measure it!!

2. Thermomiter — You could measure the degrees of the puddle. You could stick the thermomiter in the puddle to see how hot or cold it is.

3. Weight — You could measure the weight of the puddle. You could put all of the water in a bucket and measure the weight of it.

Grades 4–6

Students in the upper elementary grades communicate their thinking more clearly and with more detail. Their reports reflect the kinds of measuring devices they have been exposed to in their textbooks, but the application may still be somewhat muddled.

4. Protractor — You could take a big protractor and kind of round the puddle off to see how much of a circle the puddle is.

5. Measuring Cup — You could put all of the puddle water in a measuring cup and measure how much water is in that puddle.

6. Tangram — You could measure it with tangrams and put all of the tangrams in the puddle that it needs.

The Puddle Problem

Extending the Learning

1. As with all Puddle Questions, you'll want to follow up the assessment process by gathering the students together to share their work and the thinking they did in solving the problem. List all the different ways students thought of to measure a puddle. Include only the ways the whole group decides are practical. Make a column for tools and another for the different dimensions they planned to measure.

2. Pose this question for the students: **Of all the different ways we have listed, which do you think is the best way to measure if you wanted to compare puddles?** Divide the class into groups of four students. Have them discuss this question and come up with a decision. Hold a class discussion and let the groups debate the pros and cons of different methods.

~~ Talking It Over ~~

Which way of measuring did you decide is the best way to compare puddles?

~ We decided that depth is the most important thing to measure. Sometimes puddles are just thin but sometimes they're really deep like there's a rut or something in the road.

~ Sometimes a puddle can be deep one place and not deep somewhere else so you have to find the deepest part.

Does everyone agree that this is the best way to compare puddles?

~ No, because how are you going to find out how deep a really big puddle is?

~ I don't think so either. A big puddle can be deep. And a little puddle can be deep too. What you have to know is how big it is down and all around.

~ Our group thought perimeter was most important. And it's easiest to measure, too.

~ We decided the best way was finding the amount of water in the puddle. You can use a siphon and measuring cups. That way you know how much water is in the puddle. That's the most important thing.

~ We decided the same thing. One puddle is long. Another is deep but, you know, really small. The big thing is how much water is there.

• • •

Measurement

Follow-Up Activity

1. On a day when there are standing puddles on the playground, give the students a chance to try out their measuring plans on the real thing. First, distribute students' reports so they can recall their ideas, and give them a chance to collect the measuring tools they need to measure the puddles.

2. Outside, have each student or group of students pick a different puddle to measure. Once they're faced with measuring a real puddle, students may decide to use tools and techniques that they hadn't thought of before. Let them know that that's fine.

3. After some experimentation, have the students record the measurements they think are important for defining their puddle.

4. Back in the classroom, pose questions such as these: **Whose puddle is longest? widest? deepest? shallowest? Which one has the longest perimeter? Which has the greatest area? Which puddle has the greatest capacity?** Students should notice that the longest puddle is not necessarily the widest or deepest. **What would you need to find out in order to compare area and capacity of different puddles?**

Television Survey

Television Survey
Focus: Statistics

The first challenge in this investigation is to come up with a survey question on a topic of high general interest: television. Students compose their question and conduct a survey of the class. Then they are challenged to find an effective way to display the data and to report on what they can learn from the data.

Mathematical Ideas

- Collecting data
- Organizing and representing data
- Interpreting data

Tools and Equipment

- ◆ lined writing paper
- ◆ plain paper
- ◆ graph paper
- ◆ pencils, markers
- ◆ scissors
- ◆ calculators as needed

Grouping Arrangement

Small groups

Blackline Master in English, page 7
and in Spanish, page 8

20

Statistics

Presenting the Problem

1. Have the students work in groups to brainstorm ideas for a survey question they could ask the class about television viewing. Encourage the groups to think carefully about how the question is worded and what kind of response they want. For example, if they ask how much TV their classmates watch each day, do they want exact amounts of time? time to the nearest half hour? Will they give specific choices, or will they leave the answers open-ended?

2. Next, students must decide how to collect their data. They might write their question on a piece of paper and circulate it so their classmates can record their answers. Or they might interview every student. Allow time for the groups to collect their data. Remind the students to make their responses neat and clear.

3. Once students have collected their data, they need to study the results and find an effective way to display them. This could be a chart or graph, or some other method. **The important thing about the display is that it show your data clearly. I should be able to understand the responses you got by looking at your display.** When students have completed the display, each one should write a report giving an interpretation of the data.

Assessment Criteria

It is important to let the students know what they will be evaluated on in this assessment. Tell them that you are interested in finding out

- ✔ how they formulate their survey
- ✔ how clearly they organize and display the data
- ✔ how well they explain in writing what the data tell them

Prompts for Getting Started

Here are some things for students to think about as they work:

- What do you want to find out?
- How can you ask the question so you will get the information you want?
- How can you organize and display your information to make it clear to someone else?
- How would you interpret the information in your graph/chart?

Television Survey

Assessing the Work

In this assessment, we are interested in finding out how students organize and display information and interpret data. The students' work will probably reflect different methods of collecting and interpreting data. Their written reports may range from a simple restatement of data shown in a graph to a sophisticated interpretation of those data. In assessment, the focus is on the quality of the survey, the effectiveness of the display of data, and the mathematical thinking in the interpretation.

Questions to ask yourself while scoring a response:

- How is the survey question asked and the information gathered?
- How well are the data organized and displayed?
- Does the student analyze or draw conclusions from the data? Are the data compared?

SCORING RUBRIC

Low Response
The question asked may be too vague or too obvious to elicit reasonable data. It may be difficult to determine the data from looking at the work. There may be little or no written analysis of the data.

Medium Response
The question is reasonable. The data are organized, as in a graph, table, or list. The report may provide facts about the data, but it may not include any analysis or interpretation.

High Response
The question is well conceived. The data are well organized and clearly displayed. The report provides some analysis of the data.

Exceptional Response
The work has all the qualities of a high response and goes beyond it in some way. The data gathering, display, and/or interpretation show additional depth of mathematical thinking for the grade level. The report may include complex comparisons or analysis of the data.

Statistics

I took a poll of what people's faviorte shows are. I found out a lot of people like show I didn't think of. 50% voted other. 12% didn't have a faviorte show. 62% like show I didn't think of.

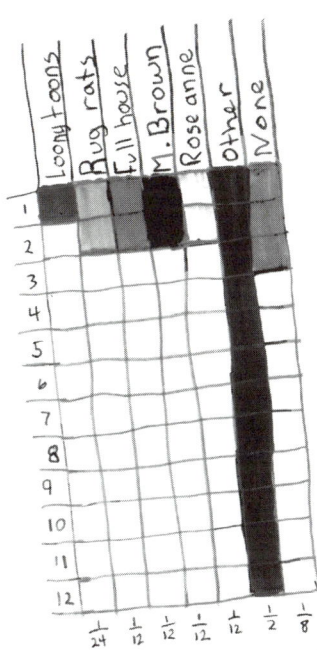

Exceptional Response

The student named several TV shows she thought would be her classmates' favorites, only to discover that they had other favorites—an interesting reflection on her polling techniques. She shows the data in both bar graph and circle graph formats, gives the fraction for each vote, and converts data to percentages.

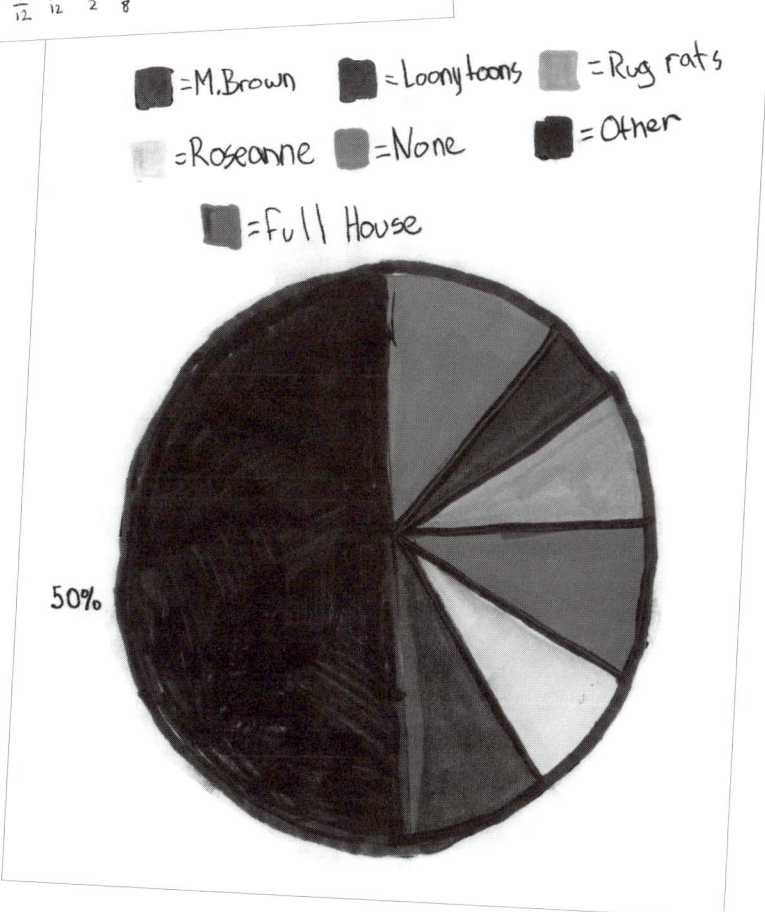

Television Survey

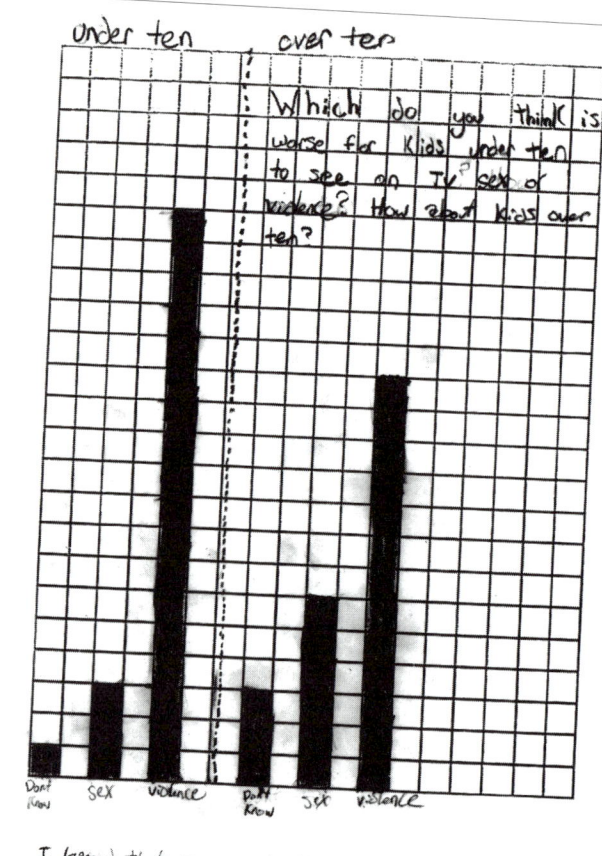

High Response

Two related questions are answered in side-by-side bar graphs in this response, and the written interpretation compares and combines the information about the answers. Both the question and the presentation make this a particularly interesting survey.

High Response

This thoughtful means of data collection works well as a data display too. It has the student respondents make a tally in a matrix in response to two questions. Unfortunately, the data collected do not lend themselves to generalizations about the relationship between the two questions.

Statistics

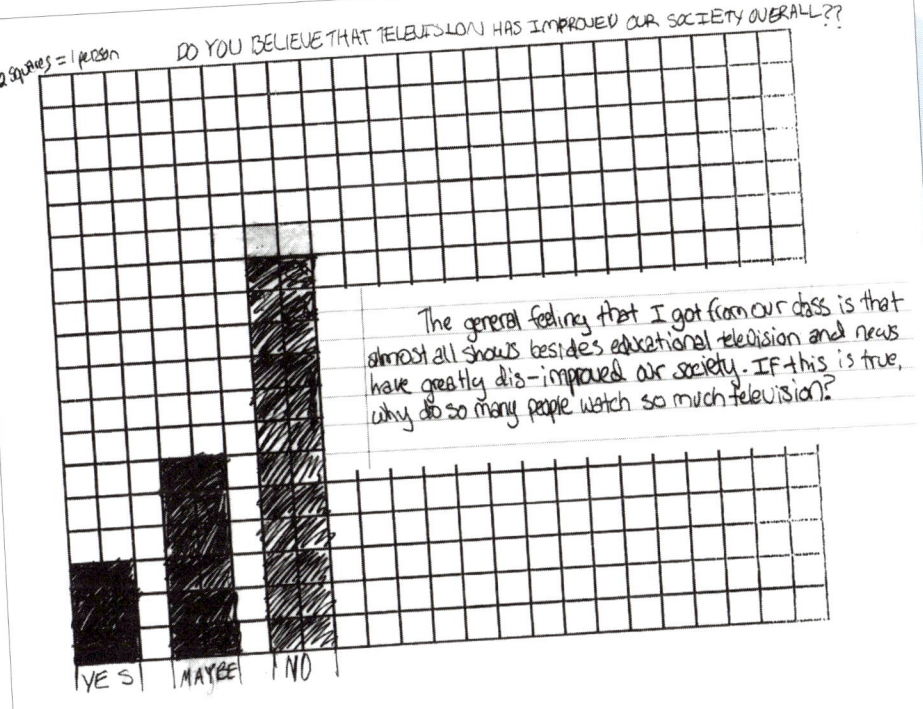

Medium Response

This student's graph is straightforward and clear. Her written report does not provide much in the way of analysis, but it does raise an interesting sociological question.

Low Response

This poll of favorite TV shows gives a list but no graphic display of data or written interpretation. Only nine students were polled.

Television Survey

Extending the Learning

1. Bring the students together and have them share the questions they used for their surveys. **How did your question work in the survey? Did you get the answers you expected? How could you change your question to make it more effective?**

2. Post or make a class set of photocopies of several different surveys that do a good a job of presenting data. Compare the different ways the data are displayed (line graphs, bar graphs, circle graphs, charts, tables, lists). Ask students to explain why they presented the data the way they did. Could they have chosen different ways?

3. For each survey, ask students to give all the information they can from the graphs or other kinds of displays. Encourage them to be specific and to back up their statements with details from the graphs.

~~ Talking It Over ~~

What did you learn about asking survey questions? How would you do it differently?

~ I'd ask a different question. Instead of Do you like TV?, I'd ask about kinds of programs that people like, or why they like certain programs. That's what people told me about anyway.

~ I'd make the question more interesting. I just asked what their favorite show was. It didn't make a good graph.

~ I put down a list of shows for people to choose from, but they didn't pick the ones I listed. They had other favorites. I guess I could have just asked people to write down their favorite show instead of pick the ones I thought of.

~ I think people should vote in secret too. If people see what other people voted for, they might vote the same way.

What did you learn about making graphs to show your data?

~ I'd make the graph tell more. Like I'd put all the people who watch one hour together and all the two hours, and I'd put them in order, from smallest time to most time. Then you could see more from just looking at the graph.

~ If the graph is good, you can write more, too. Like most people are grouped around 1 or 2 hours a day, then there's less and less people down to no TV and up to 3 hours a day.

• • •

Statistics

Follow-Up Activity

1. Tell students they will extend their survey to another class and compare the data. **First get together with your group again and look at your question. You can keep the same question, or change the question a little, or make a new question. Then you can choose the class you would like to survey.** Be sure students understand that to compare survey data, they must ask both classes exactly the same question. Also, ask them to think about the other class they choose. Do they want students of their age or a different age? If you can arrange it, your students might even survey a class at another school—high school or middle school students, for example.

2. If the groups have a revised or new question, they will begin by surveying their own class again; then they will prepare materials to survey the second class.

3. When all the data have been collected, have students work in their groups again to decide how to display the information.

4. Have groups present their reports on the new data. Encourage students to compare the two classes' results and to discuss what they can learn from the combined data. You might arrange for your students to report their results to the other classes they surveyed.

27

One to a Million

Focus: Estimation and Reasoning

INVESTIGATION 3

In this investigation, we take a look at students' reasoning skills as they work on the task of estimating how long it will take to count to a million. The problem gives us a view of what students know about place value and time relationships. As with all Puddle Questions, there are many ways to approach the problem and many levels of difficulty at which to solve it. By the end of their exploration, students may have a stronger sense of what a million is.

Mathematical Ideas

- Making estimates
- Using number sense
- Working with time
- Using multiplication and division
- Reasoning logically

Tools and Equipment

◆ blank paper
◆ lined writing paper
◆ pencils
◆ calculators
◆ clock or watch with a second hand

Grouping Arrangement

Small groups or individuals

Blackline Master in English, page 81 and in Spanish, page 82

Estimation and Reasoning

Presenting the Problem

1. Give each student a copy of the investigation. **Think about how long it might take you to count to one million. How might you go about making a reasonable estimate? What are some things to consider as you tackle the problem?**

2. Let the students talk together in small groups about how to approach the problem. They might want to work together to time each other in their counting.

3. Have students write individual reports giving their estimates and explaining their thinking. **Be sure to tell exactly how you came up with your estimate. Show all of your figuring.** Be sure calculators are available to students who want to use them.

Assessment Criteria

It is important to let the students know what they will be evaluated on in this assessment. Tell them that you are interested in finding out

- ✔ what strategy they use for making an estimate
- ✔ whether their estimates are based on a reasonable approach
- ✔ how clearly they show how they arrived at their estimates

and to a lesser degree

- ✔ whether their computation is correct

Prompts for Getting Started

While the students work on the problem, circulate among them and observe the approaches they are taking. Pose questions like these:

- What are some different ways you could approach the problem?
- How can you make your estimate as accurate as possible?
- How can you show your thinking so everyone will understand your reasoning?

One to a Million

Assessing the Work

With this assessment question, we are interested in finding out how well students are able to take a logical approach to making an estimate. Their answers may give you some insights into their understanding of the numeration system and their facility in converting units of time (minutes to hours or days, for example). Whatever their approach to solving the problem, the answer is far less important than the method they use to arrive at it. You may want to have a calculator handy to check students' figures and follow their reasoning.

Questions to ask yourself while scoring a response:

- What strategy is the student using to make an estimate?
- Is the reasoning logical?
- How clear is the explanation of how the estimate was determined?
- Are there any computational errors or misconceptions?

SCORING RUBRIC

Low Response
The estimate is based on guessing, or the reasoning is completely flawed. Communication is limited.

Medium Response
Some thoughtful reasoning is revealed in the report, although the reasoning may be flawed or incomplete. Communication is fair.

High Response
The estimate is based on logical and thoughtful reasoning, and the response indicates some understanding of the complexity of the problem. Estimates are given in reasonable terms, for the most part. The explanation is clear and fairly detailed.

Exceptional Response
The response reveals exceptional logical reasoning. The estimate is based on complex calculations that show a high level of thinking about the question. Computations are reasonably correct. Communication is excellent.

Estimation and Reasoning

About how long does it take you to count to one million?

The first thing I did was find out how many seconds it takes to count each digit. I got 1 digit 1 sec, 2 digits 1 sec, 3 digits 3 sec, 4 digits 5 sec, 5 digits 6 sec, 6 digits 7 sec, 7 digit 2 sec. Next I found out how many 7 digit numbers there are. Then when I found out how many numbers there in each digit I multiplied by how many seconds it took to say each digit. In all I added up all the numbers and got an answer wich is 6,887,799 ~~minutes~~ seconds. That is about 79 days.

9	1 digit	9 sec
90	2 digit	90 sec
900	3 digit	2700 sec
9,000	4 digit	45,000 sec
90,000	5 digits	540,000 sec
900,000	6 digit	6,300,000 sec
		6,887,799 sec

6,887,799 ÷ 60 (sec) ÷ 60 (min) ÷ 24 (hours) = 79.719895

Exceptional Response

Estimate: 79 days. This student reveals an exceptionally strong understanding of place value. She takes into account how much longer it takes to say multiple-digit numbers than single-digit numbers. She also figures out how many one- to six-digit numbers there are in one million (9 one-digit numbers, 90 two-digit numbers, 900 three-digit numbers, and so on).

High Response

Estimate: 13 days. This student timed how long it took to count to 1,000—18 minutes. She used that rate to figure how long it would take to count to one million, a solid approach even though it does not consider lengthening time for counting numbers above 1,000. Her conclusion is insightful—estimates will vary depending on how fast people count.

About how long would it take you to count to one million?

I think it would take about 12 to 13 days to count to one million, because when I timed myself counting to 1,000, it took me exactly 18 mins. Then I divided 1,000,000 by 1,000. I got 1,000. I then mulitplyed 1,000 by 18 mins. I got 18,000 mins. I divided 18,000 mins by 60 mins, since there are 60 mins in 1 hr. That was 300 hrs. I divided 300 hrs by 24, since there are 24 hours in a day. My answer was 12½ days. I rounded that to 13 days. So it would take you about 13 days to count 1,000,000. If you count to 1,000 in 18 mins. If you count faster than me, then your answer will be less. If you count slower than me, your answer will be more.

31

One to a Million

High Response

Estimate: 6 days and 6 hours. In this report every step is spelled out, with the figuring included. The approach is good; group members timed themselves and averaged how long it took them to count to 100. The average was used to calculate the time required to count to a million. This answer was "transformed into more understandable figures."

My group decided that everyone in our group should count to 100 and then work from there.

Person	Time (in sec.)
Caroline	61 sec.
Sarah	54 sec.
Jamie	53 sec.
Maura	47 sec.

Then we found the average of the group.

```
  54         53.9 (rounded) → 54
  53       4)215
  61         20
+ 47         15
 215         12
              3
```

Then we figured out there was a ten thousand hundreds in one million.

```
        10,000
100)1,000,000
```

We then multiply 10,000 by 54 to find how much time it would take to one million.

```
   10,000
 ×     54
   40000
  500000
  540,000 sec.
```

I found out how many seconds were in an hour.
60 sec. in a min., 60 min. in an hour.

```
    60
  × 60
    00
  3600
  3,600 hours
```

So then we transformed 540,000 sec. into a more understandable figures of numbers. We divided 3,600 into 540,000 to find how many hours it took us to count to one million.

```
          150 hours
3,600)540,000
       3,600
       1,800 0
       18000
          00
         - 0
           0
```

Then we said there are 24 hours in a day. Then we divided 24 hours into 150 hours to find the total answer.

```
      6.6
  24)150
     144
     006
```

It would take us 6 days and 6 hours to count to one million.

Estimation and Reasoning

> I counted to 500 it took me 7 minutes I multiplied it by 2 because I wanted to know how long it would take to count to 1,000 because then I could times it by 1,000 because it would come out as the answer.
>
> ```
>      14
>  x 1000
>  -----
>   4000
>  10000
>  -----
>  14,000
> ```
> Hours it would take for me to count to 1,000,000

Medium Response

Estimate: 14,000 hours. After counting to 500, this student doubled the time and multiplied the product by 1,000. This gives the minutes (not hours, as she states) it would take to count to a million. Though the response shows some creditable thinking, the reasoning is flawed; and the hours figure of 14,000 is not converted to a more understandable unit.

> 100 X 10 = 100,000
> It took me 1 minute to count to 100 so 10 x 100 makes it 10-minuets to count to 1,000,000.

Low Response

Estimate: 10 minutes. This student started with a good strategy by finding how long it takes him to count to 100. He errs in thinking that there are ten rather than ten thousand 100s in one million. It's significant that he did not question his results and wonder how he could possibly count to such a high number in only ten minutes.

One to a Million

Extending the Learning

1. Record on the chalkboard, in random order, all of the estimates students made for the amount of time it would take them to count to a million. Then ask, **What are some things you notice about the estimates?**

2. Talk about the different ways the students arrived at their estimates. If they used different estimates for different number spans, have them explain their thinking to the class.

3. Organize the students into pairs and have them exchange reports. Let the students give each other feedback on whether their reasoning makes sense and whether their reports communicate their thinking clearly.

~~ Talking It Over ~~

What are some things you notice about our estimates?

~ Some are in hours, some are in days. There's one in seconds.

Does the unit make a difference?

~ The numbers are bigger for seconds than they are for days.

~ If you tell how many seconds it takes to count to a million, the number is so big you can't really know what it means. But with days, you know.

~ You need to give your estimate in an understandable form.

What are some things you took into account when you made your estimate?

~ You might not want to count every day all day long. You might need to sleep. So I made an estimate for a 24-hour day. And I made another estimate for a 10-hour day.

~ Another thing is that people count at different speeds.

~ Yeah. We took an average in our group.

~ As you count higher it takes longer to say the numbers. One- and two-digit numbers take 1 second. Three- and four-digit numbers take 2 seconds. Five- and six-digit numbers take 4 seconds. That's how I got my estimate.

• • •

Estimation and Reasoning

Follow-Up Activity

1. Have the students work in small groups to analyze the problem further by studying the number one million and figuring out how many one- to seven-digit numbers there are in one million. Their findings can be presented in a chart, diagram, or table. Example:

Digits	Numbers	How many?
1	1–9	9
2	10–99	90
3	100–999	900
4	1,000–9,999	9,000
5	10,000–99,999	90,000
6	100,000–999,999	900,000
7	1,000,000	1

2. Have the students time how long it takes them to count each type of number and add those figures to their reports. Finally, they may wish to use these figures to calculate a revised estimate for the length of time it would take to count to a million.

Equations Galore!

Equations Galore!
Focus: Arithmetic Operations

In this task, students explore equations for $\frac{1}{2}$. There is clearly no one right answer to this problem, for the number of possible equations is limitless. What is interesting is how students tackle the problem, how they organize their thinking, and what kinds of equations they are comfortable listing. The results can be revealing in a number of ways. Students may not list equations you feel they have had lots of exposure to. Error patterns can suggest misunderstandings you may never have suspected. Or, students may list equations that are more sophisticated than you expected.

Mathematical Ideas

- Understanding operations
- Understanding fractions
- Using patterns
- Making organized lists

Tools and Equipment

◆ blank paper
◆ lined writing paper
◆ pencils
◆ fraction models
◆ calculators

Grouping Arrangement

Individuals

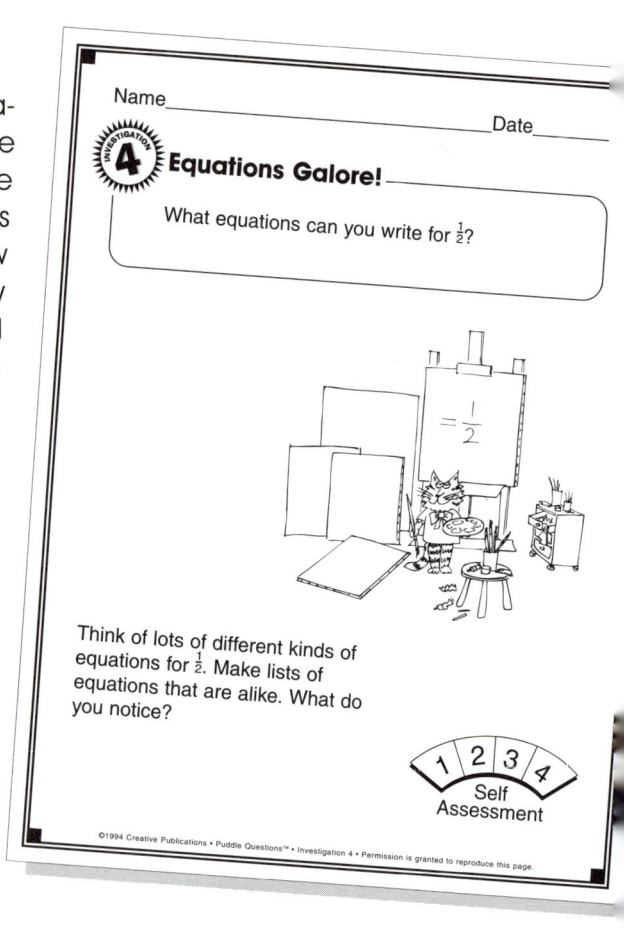

Equations Galore!

What equations can you write for $\frac{1}{2}$?

Think of lots of different kinds of equations for $\frac{1}{2}$. Make lists of equations that are alike. What do you notice?

Blackline Master in English, page 8
and in Spanish, page 8

Arithmetic Operations

Presenting the Problem

1. Start by asking the students what an equation is. List equations students suggest on the chalkboard.

2. Explain that today the students will be exploring equations for $\frac{1}{2}$. Write $= \frac{1}{2}$ on the chalkboard and say, **Each of the statements you write should equal $\frac{1}{2}$. Think of all the different kinds of equations you can write for $\frac{1}{2}$. Make a list for each different kind.**

3. Students may wish to use fraction models to help them come up with ideas for equations. If you have fraction models and the students are familiar with them, make them available. Students should also have access to calculators if they wish to use them.

4. Students may begin by randomly listing equations. Once they see some patterns, they may decide to start on a fresh sheet of paper and organize their equations in some way. Remind them that it is their responsibility to show what they know in these assessment tasks. Encourage them to do their best work.

Assessment Criteria

It is important to let students know what they will be evaluated on in this assessment. Tell them that you are interested in finding out

✔ how many different kinds of equations they can write
✔ whether they organize their equations in any way

and to a lesser degree

✔ whether their equations are correct

Prompts for Getting Started

You may want to ask questions like these as students are working on the problem:

- **Can you think of another kind of equation?**
- **Could you write more equations like the ones you've already written? How?**
- **What are some especially interesting equations you can write?**

Equations Galore!

Assessing the Work

In this assessment, we are interested in finding out what different ways students are able to construct equations, what operations they are comfortable using, any misconceptions they may have about fractions, and their understanding of the relationships among like equations. Your students may write equations using addition, subtraction, multiplication, division, or mixed operations with fractions and/or whole numbers. The equations may be organized by type (all the addition equations together, for example), or they may be listed in a special way to show a pattern (for example, $\frac{1}{4} + \frac{1}{4} = \frac{1}{2}, \frac{1}{8} + \frac{1}{8} + \frac{1}{8} + \frac{1}{8} = \frac{1}{2} \ldots$; or $\frac{2}{4} = \frac{1}{2}, \frac{3}{6} = \frac{1}{2}, \frac{4}{8} = \frac{1}{2} \ldots$).

Questions to ask yourself while scoring a response:

- How many different kinds of equations did the student list? Did the student list any creative or unusual equations?
- Did the student organize the equations in any way? Did the student use patterns in listing equations?
- Are any error patterns or misconceptions revealed in the equations?

SCORING RUBRIC

Low Response
There is little response to the problem. The equations may not be organized in any perceivable way and may reveal significant errors.

Medium Response
The response is acceptable, but limited in the number of different kinds of equations that are listed. The equations may or may not be organized clearly, and they may reveal a significant pattern of errors or a major misconception.

High Response
The response is substantial. Several different kinds of equations are listed, and they are probably organized clearly. There are few, if any, significant errors. The student may have made use of patterns to extend the number of equations in a list.

Exceptional Response
The response has all of the qualities of a high response, and it is extraordinary in some way. For example, the student may have listed especially interesting, creative, or sophisticated equations that indicate a higher level of mathematical thinking.

Arithmetic Operations

math technique used	example	$\frac{1}{2}$
percent	50% =	$\frac{1}{2}$
adding percents	25% + 25% = 50% or	$\frac{1}{2}$
subtracting percents	75% − 25% = 50% or	$\frac{1}{2}$
multiplying percents	25% × 2 = 50% or	$\frac{1}{2}$
dividing percents	100% ÷ 2 = 50% or	$\frac{1}{2}$
fractions	$\frac{2}{4}$ =	$\frac{1}{2}$
adding fractions	$\frac{1}{4} + \frac{1}{4} = \frac{2}{4}$ or	$\frac{1}{2}$
subtracting fractions	$\frac{5}{6} - \frac{2}{6} = \frac{3}{6}$ or	$\frac{1}{2}$
multiplying fractions	$\frac{2}{8} \times 2 = \frac{4}{8}$ or	$\frac{1}{2}$
dividing fractions	$\frac{8}{8} \div 2 = \frac{4}{8}$ or	$\frac{1}{2}$

Exceptional Response

Starting with a rough draft, this student first listed equations more randomly, then organized her ideas on a new sheet of paper in the chart you see here. She organized her paper exceptionally well. She applies all four operations to both fractions and percentage values and gives an example of each type. The assumption is that there are many more she can list for each type.

Exceptional Response

This student's work indicates a fluency with all the operations, an understanding of how fractional equations relate, and an interest in exploring patterns in equations. He found an interesting pattern in the series of equations
$2 \div 2 \div 2 = \frac{1}{2}$,
$4 \div 2 \div 2 \div 2 = \frac{1}{2}$, and on to $128 \div 2 \div 2 \div 2 \div 2 \div 2 \div 2 \div 2 \div 2 = \frac{1}{2}$.

$1 \div 2 = 1/2$
$2 \div 2 \div 2 = 1/2$
$4 \div 2 \div 2 \div 2 = 1/2$
$4 \div 4 \div 2 = 1/2$
$1 + 1 \div 2 \div 2 = 1/2$
$2 + 2 \div 2 \div 2 = 1/2$
$1/2 \times 2 \div 2 = 1/2$
$16 \div 2 \div 2 \div 2 \div 2 \div 2 = 1/2$
$32 \div 2 \div 2 \div 2 \div 2 \div 2 \div 2 = 1/2$
$128 \div 2 \div 2 \div 2 \div 2 \div 2 \div 2 \div 2 \div 2 = 1/2$
$4000000 \div 4000000 \div 2 = 1/2$
$4 \div 8 = 1/2$
$10 \div 20 = 1/2$
$200 \div 400 = 1/2$
$300 \div 600 = 1/2$
$600 \div 1200 = 1/2$
$100 \div 200 = 1/2$
$5 \times 5 - 24 1/2 = 1/2$
$1/4 + 1/4 = 1/2$
$1/4 \times 2 = 1/2$
$1/8 \times 4 = 1/2$
$1/16 \times 8 = 1/2$
$1 - 1 + 1/2 = 1/2$
$1/2 + 1/2 - 1/2 = 1/2$
$1 - 1/2 = 1/2$
$300 - 299 1/2 = 1/2$
$300 - 300 \div 1200 = 1/2$

Equations Galore!

High Response

This response is quite substantial and well organized, though limited to addition equations for $\frac{1}{2}$. The student listed a series of equivalent fractions for $\frac{1}{2}$ down the left side of the paper, and then to the right provided a list of many of the corresponding addition equations for each different denominator. She has explored one idea thoroughly.

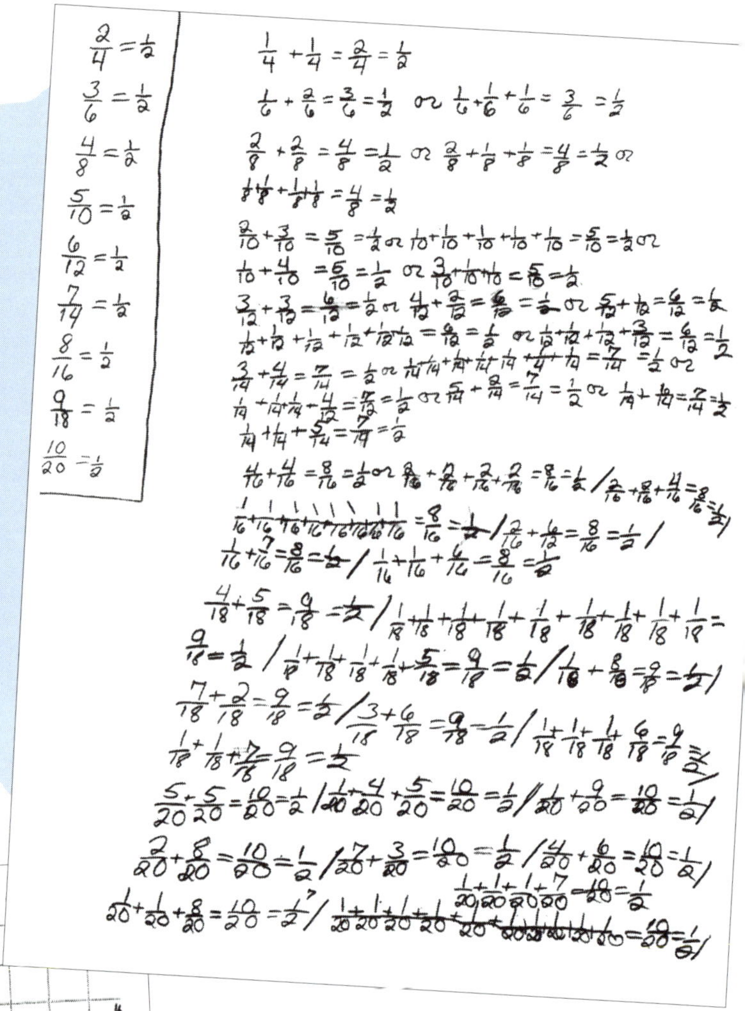

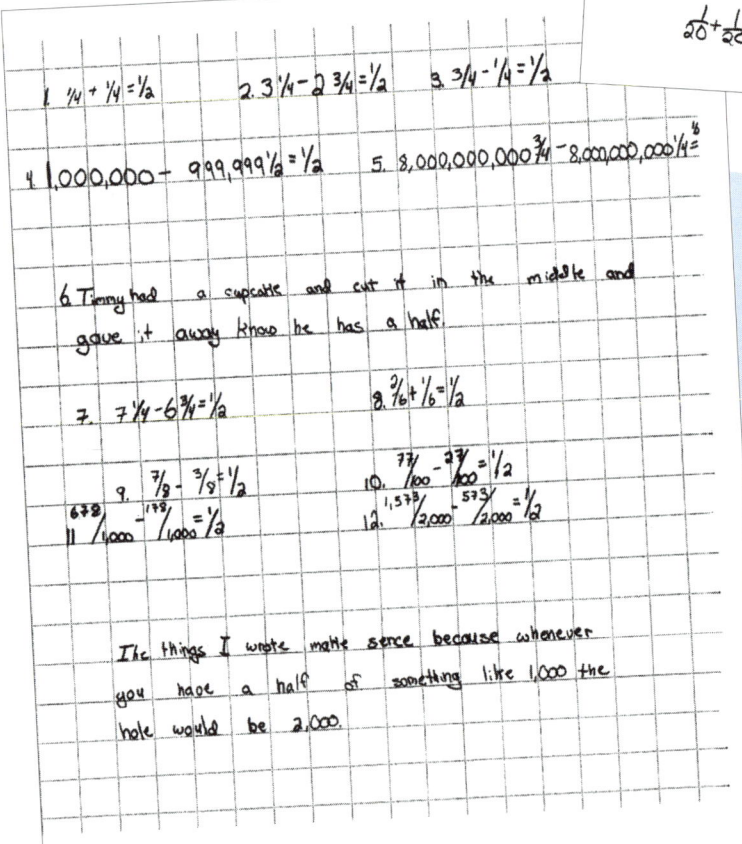

High Response

Some challenging addition and subtraction equations with fractions and whole numbers are included in this student's work. He also explained how he could work backward to find the whole amount if given a certain value for a half.

Arithmetic Operations

Student Work 1

What equations can you write for 1/2?

1. $0.6 - 0.1 = 0.5$
2. $0.7 - 0.2 = 0.5$
3. $0.8 - 0.3 = 0.5$
4. $0.9 - 0.4 = 0.5$
5. $0.10 - 0.5 = 0.5$
6. $0.11 - 0.6 = 0.5$
7. $0.12 - 0.7 = 0.5$
8. $0.13 - 0.8 = 0.5$
9. $0.14 - 0.9 = 0.5$
10. $0.15 - 0.10 = 0.5$
11. $0.16 - 0.11 = 0.5$
12. $0.17 - 0.12 = 0.5$
13. $0.18 - 0.13 = 0.5$
14. $0.19 - 0.14 = 0.5$
15. $0.20 - 0.15 = 0.5$
16. $0.21 - 0.16 = 0.5$
17. $0.22 - 0.17 = 0.5$
18. $0.23 - 0.18 = 0.5$
19. $0.24 - 0.19 = 0.5$
20. $0.25 - 0.20 = 0.5$

etc. etc. etc.

What do you notice? Well first I am telling you something that you might or might not know. It's that 0.5 is the same as 1/2. Now I notice that when I first subtracted 0.6 and 0.1 you get 0.5. Then I thought that if I add 0.1 to 0.6 and 0.1 to 0.1 the total of the two number will 0.7 and 0.2. Well if I subtract 0.7 and 0.2 I will get 0.5. My point is that if I continue adding 0.1 to 0.7 and 0.2 and their totals I will get 0.5 a lot of times like this:

$0.6 - 0.1 = 0.5$
$+0.1 \quad +0.1$
$0.7 - 0.2 = 0.5$
$+0.1 \quad +0.1$
$0.8 - 0.3 = 0.5$

and on. etc

Medium Response

This student explores one equation type, using a subtraction pattern. The pattern is nicely described, and if he had done what he said, all of his equations would be correct. In number 5, when he added 0.1 to 0.9 he got 0.10 instead of 1.0. From then on, he added 0.01 instead of 0.1, creating a consistent pattern of errors in his $\frac{1}{2}$ equations.

Student Work 2

1.) $1 - 1/2 = 1/2$

2.) $0 + 1/2 = 1/2$

3.) $1 \overline{)1/2}^{1/2}$

4.) $1\frac{1}{2} - 1 = 1/2$

1. One minus 1/2 equals 1/2.
2. Zero plus 1/2 equals 1/2.
3. 1/2 divided by 1 equals 1/2.
4. One and 1/2 minus one equals 1/2.

extra- you could always write out your problems like I did for # 1,2,3,4.

Low Response

Here the student lists only four equations, one of which is incorrect. She does a good job of writing the equations with words, but does not explain how she arrived at her equations.

Equations Galore!

Extending the Learning

1. Have the students pair up with a partner and share their papers. Partners should each take a turn showing what equations they listed and explaining any discoveries they made. Students might also talk about what self-assessment score they gave themselves and why.

2. Bring the whole class together to discuss what they found out. **What kinds of equations did your partner list that you did not? What did you include that your partner did not?**

3. Have the students discuss what they learned in this investigation. Talk about the different kinds of equations and any patterns they found.

~~ Talking It Over ~~

What did you notice as you were working on your equations for this investigation?

~ I noticed that all of them equal $\frac{1}{2}$, and that it would take a long time before you could find all of them.

~ I noticed that too. There are a billion ways to get $\frac{1}{2}$.

~ Yeah, the equations will go on forever. So will fractions. As long as there are numbers there will be fractions and equations. Like, $10 \div 20 = \frac{1}{2}$ so that means $20 \div 40 = \frac{1}{2}$ and $30 \div 60 = \frac{1}{2}$. You double for the second number but go up 10 for the first number.

Did anyone else find patterns like that one that go on forever?

~ You can take a number and add one half to it and just subtract the number you used. Like $40 + \frac{1}{2} - 40 = \frac{1}{2}$. Or $50 + \frac{1}{2} - 50 = \frac{1}{2}$.

~ I noticed that when I first subtracted 0.6 and 0.1, you get 0.5. You might not know it, but 0.5 is the same as $\frac{1}{2}$. So I thought if I add 0.1 to 0.6 and 0.1 to 0.1 the total of the two numbers will be 0.7 and 0.2 and I will get 0.5 again. My point is that if I continue adding 0.1 to both numbers, I'll get 0.5 a lot of times.

~ Here's another pattern: Every time you can multiply the bottom number by 2, you multiply the number on the outside by 2. So you get $\frac{1}{4} \times 2$, $\frac{1}{8} \times 4$, $\frac{1}{16} \times 8$. Like that. That'll always equal $\frac{1}{2}$.

• • •

Arithmetic Operations

Follow-Up Activity

Equations for $\frac{1}{2}$ shown on poster:
- $16 \div 32 = \frac{1}{2}$
- $2 \div 4 = \frac{1}{2}$
- $20 \div 40 = \frac{1}{2}$ (partially covered)
- $6 \div 12 = \frac{1}{2}$
- $22 \div 44 = \frac{1}{2}$
- $8 \div 16 = \frac{1}{2}$
- $24 \div 48 = \frac{1}{2}$
- $14 \div 28 = \frac{1}{2}$

1. Have students work in pairs on this activity. Explain that they will make posters to show a set of equations that follow a single pattern. Then they will cover some of the equations to make a puzzle for their classmates. **Look through the lists of equations each of you wrote for this investigation and pick out a set that follows a pattern. You'll need about ten to fifteen equations in your set.**

2. Hand out large pieces of poster paper for the students to use in preparing their Pattern Puzzle Posters. Have students write the equations on the poster, making them big enough to be seen easily by the rest of the class. Then students will choose some of the equations to cover up with strips of paper. They should leave enough equations showing so that their classmates might still find a pattern, but cover enough to make it a challenge. The paper strips can be taped in place on the top side to allow for looking underneath later on.

3. Display a different pair's poster each day and challenge the rest of the class to identify the missing equations.

Doghouse Designs
Focus: Design and Construction

INVESTIGATION 5

In this investigation, students tackle an architectural problem, and in the process explore different ways that flat surfaces can be put together to form a three-dimensional solid. This requires strong spatial and visual thinking skills. Students approach the task of creating their own doghouse designs in many interesting ways.

Mathematical Ideas
- Describing geometric shapes
- Measuring geometric shapes
- Thinking visually
- Using spatial sense
- Drawing to scale
- Estimating

Tools and Equipment
- lined writing paper
- blank paper
- graph paper
- dot paper
- pencils
- rulers
- scissors
- tape

Grouping Arrangement
Small groups or individuals

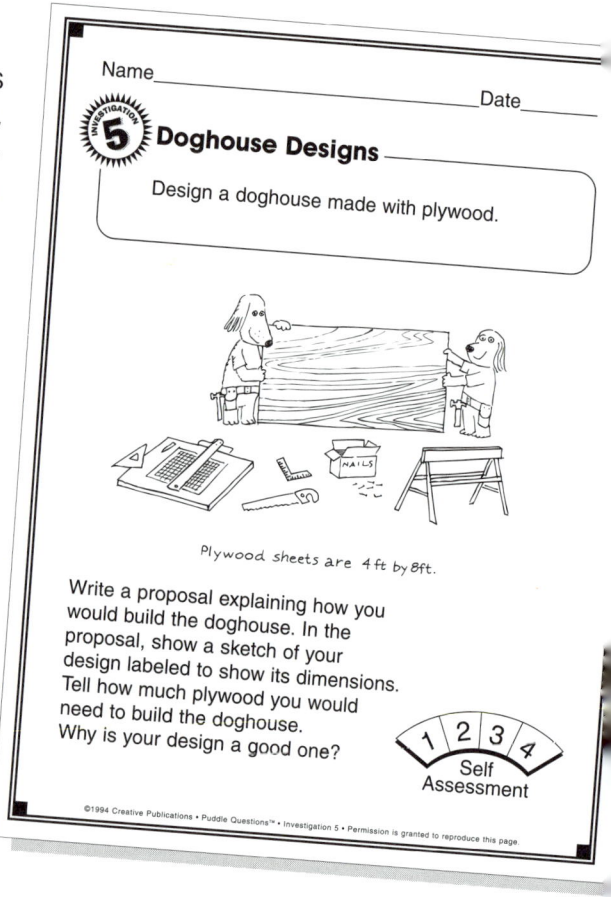

Blackline Master in English, page 8
and in Spanish, page 8

Design and Construction

Presenting the Problem

1. Explain to the students that their task is to design a doghouse made from plywood sheets that measure 4' × 8'.

2. Give students access to graph and/or dot paper, tape, scissors, and rulers so they can try out different design ideas. When they think they have a good design, have them prepare their final design proposal. This should include sketches of the pieces, labeled to show their measurements, and a written description of the design.

3. Tell students that they will have a chance to build cardboard models of their doghouse designs later on. For now, they should concentrate on creating the design proposal, not on a finished product.

Assessment Criteria

It is important to let students know what they will be evaluated on in this assessment. Tell them that you are interested in finding out

- ✔ whether they can create a workable design for a doghouse
- ✔ whether they can figure out how much plywood they will need to make the doghouse
- ✔ how clearly they describe their design and how to build the doghouse from the design

Prompts for Getting Started

Students who have difficulty deciding how to tackle this task could be prompted with questions like these:

- How will you decide what size to make your doghouse?
- What different shapes could your doghouse be?
- How can you show how you would cut the plywood to make the doghouse?
- How can you show how many pieces of plywood you will need?

Doghouse Designs

Assessing the Work

You may be surprised at which students excel at this investigation. Students gifted with spatial sense are often not the same ones who excel at other mathematical challenges. While the students may create doghouse designs with elaborate extras (such as an automatic feeding system), for assessment the focus is on the geometry and measurement skills revealed in the response.

Questions to ask yourself while scoring a response:

- Does the design work to produce a complete doghouse? Do the pieces fit together as described in the proposal?
- Did the student accurately determine the amount of plywood needed for the project?
- How clearly does the student describe the design and how to build it? Are the steps in the construction process easy to follow?

SCORING RUBRIC

Low Response
The student has had little success producing a workable design for a doghouse. The amount of plywood needed, if given, may not match details of the design. The report may not be clear or detailed.

Medium Response
An effort has been made to create a workable design, but there may be significant flaws in its conception. The report is fairly clear, with some detail about the construction, but the description may not be complete.

High Response
The design is a workable plan. For the most part, the pieces fit together as designed. The steps in the construction process are fairly clearly described.

Exceptional Response
The design has all the qualities of a high response, and it shows exceptional mathematical thinking for the grade level. The design may be unusually complex or detailed. Or, it may employ advanced geometric ideas or visual thinking skills.

Design and Construction

For this dog house you need 5 sheats of plywood — 2 sheats for the roof, 1 sheat (cut) for the sides and 1 sheat (cut) for the front and 1 sheat (cut) for the back.

For the front, you need to cut a shape similar to a triangle (figure 1) then cut a hole in the middle for the door. For the back cut another shape that is the same as the front but with no door (figure 2). For the sides take only 1 sheat of plywood and cut two peaces that measure 1' by 8', and for the back use two whole peaces for the roof. To asemble the house nail the two sides to the front and back. Now put the roof on by nailing the roof peices to the sloped side of the house (figure 6). This house is a good disine because the roof is sloped so it sheads rain and it wont be nocked over by the wind because there is no over hang for the wind to go under.

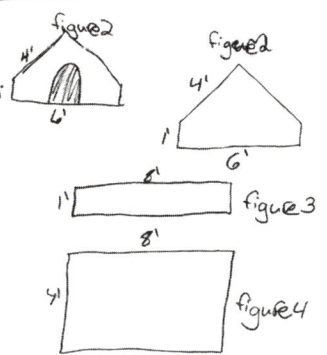

Exceptional Response

This doghouse has a high peaked roof with low outside walls. The construction process is clearly explained through the written report and diagram. The student has correctly totaled the plywood needed and has justified her answer.

Exceptional Response

This student went beyond the requirements of the task to produce a doghouse and fenced dog yard. The doghouse is a cube made of six squares cut from three boards. The fence rails are made of 1' × 8' boards and 1' × 4' boards cut from three more sheets of plywood.

1. Buy six 8'×4' plywood boards, but take three of them, cut them in half so they are all six even squares, so they should be 4'×4'.
2. Put one square on the ground which is the bottom and nail four walls on to it.
3. Now that the walls are on lay the top of the house (or the roof) on the top and nail it in. Now you have the main house.
4. With a chain saw cut two doors, one on the front side and the other on the side right of the front. The doors should be 2'×2'.
5. Buy twelve 5 ft tall posts for your fence: four for the eating yard, five for the play yard & four for the exercize yard. (Look at picture above)
6. The play yard and exercize yard should have a gap in it so your dog can go back and forth.
7. With the rest of the 8'×4' plywood boards, cut ten 1'×8' rails for inbetween the posts of the outside rails.
8. With the remaining plywood which should be 2'×8' cut four 1'×4' rails for the inside. Between the eating and play yards there should be one, between the play and exercize yard there should be three, look at picture.

47

Doghouse Designs

High Response

This student includes sketches of the top and side views of the doghouse and diagrams of how the plywood sheets should be cut. His report, somewhat difficult to follow, contains some minor conceptual flaws, but the end result is a workable design.

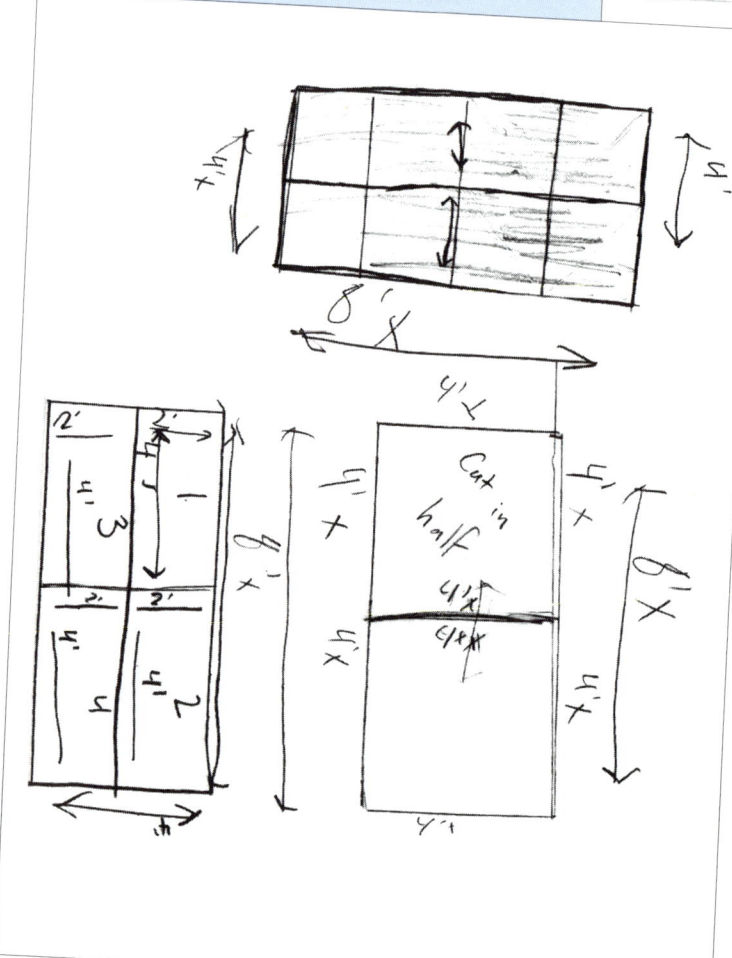

Proposal

First I cut one of the pieces in half long ways. These are the two long walls. Then I cut another piece in four long quarters. I put two of them for the front and back walls. Then I can cut the long walls in half the other way so they match the roof, cut like this. So that both sides of the roof are equal.

On the front and back I need something to hold up the roof. So cut two pieces looking like this. So that the roof can slant and rain can't make it collaps.

I have to put a door on it so on the front short wall I cut a door out. Then I mount it on one big piece of plywood. Look at next page to see how it'll turn out.

You should buy this dog house because it has a big overhang so rain'll come off easily. It has a 2 foot overhang and is 4 feet wide, 4 feet long and the roofs peak is also 4 feet. The walls are two feet high. They are perfect for any normal sized dog. The detailed sketch is on the next page.

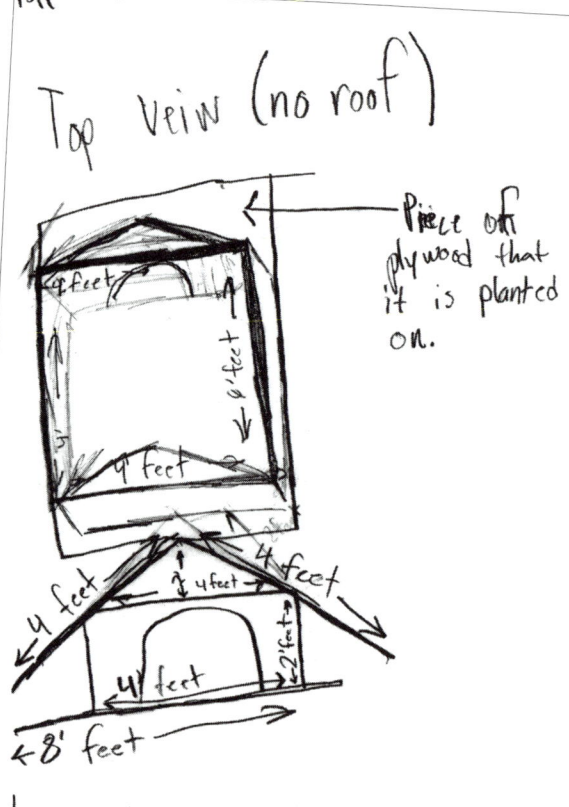

Top veiw (no roof) — Piece of plywood that it is planted on.

Side veiw

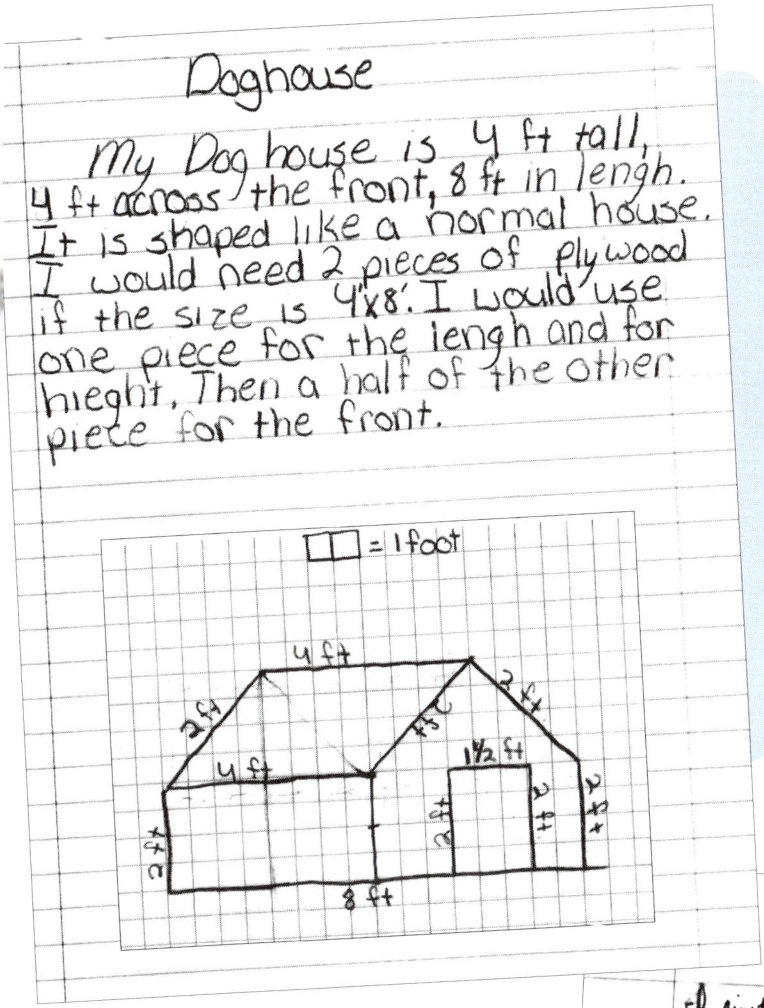

Design and Construction

> **Doghouse**
>
> My Dog house is 4 ft tall, 4 ft across the front, 8 ft in length. It is shaped like a normal house. I would need 2 pieces of plywood if the size is 4'x8'. I would use one piece for the length and for hieght, Then a half of the other piece for the front.

Medium Response

This doghouse is beautifully drawn and labeled, with the horizontal and vertical lines drawn to scale. However, the drawing does not take depth into consideration. If you were to cut out the pieces, you would discover that they can't fit together in a three-dimensional model.

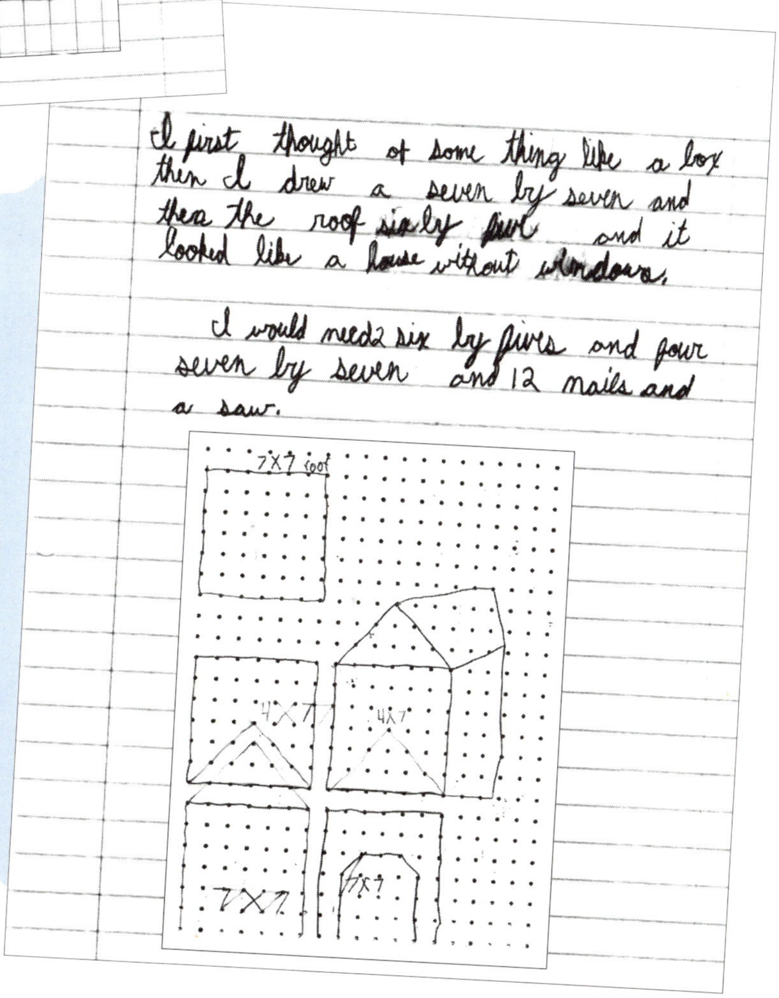

> I first thought of some thing like a box then I drew a seven by seven and then the roof six by four and it looked like a house without windows.
>
> I would need 2 six by fives and four seven by seven and 12 nails and a saw.

Low Response

The dimensions and number of plywood pieces are given in this report, but how they fit together to make a doghouse is unclear.

Doghouse Designs

Extending the Learning

1. Have the students pair up and share their designs with a partner. They should take turns explaining how they designed their doghouse and why they think their design is a good one. Then have the students compare their designs. **How are your doghouses alike and different?**

2. As a whole class, talk about the different sizes of doghouses that students came up with. **How did you decide what size to make your doghouse? How did the size of available plywood affect your design?**

~~ Talking It Over ~~

How did you decide what size to make your doghouse?

~ I thought about my dog. He's about 2 feet tall, I think. I figured the doghouse should be about 3 feet tall so my dog won't bump his head on the ceiling when he stands up.

~ Yeah, I thought my doghouse should be $2\frac{1}{2}$ feet wide in case my dog gets fat.

~ I thought about the size, and I decided to make it small and simple because it's just for a dog to sleep in. So I decided to make it the same size as my dog's bed.

How did the size of the plywood affect your design?

~ First I thought about the plywood and how many pieces I could get out of one sheet.

~ One piece isn't very much. If you use just one piece, you'd have to have a really small dog.

~ I used a whole bunch of sheets—six of them. I didn't cut them at all.

~ I cut mine because I thought that a 4 by 8 piece of plywood is kinda tall for a wall. I took one piece and cut that. I cut them again so I got four pieces of wood that were 4 feet by 2 feet. That seemed to be some good sizes to work with.

• • •

Design and Construction

Follow-Up Activity

1. Tell students they will now build scale models of the doghouses they have designed. Discuss the idea of scale, asking students for ways to approach this challenge. They will want to choose an appropriate scale for building, as well as one that allows them to make accurate and understandable calculations. They should write the scale they choose on their original plan.

2. Make calculators available, as well as tagboard, graph paper, scissors, tape, and other materials of your choice, and have students create their models. As the students work, they may discover flaws in their designs and decide to revise their doghouse proposals.

3. Arrange a place for students to display their models. For an additional challenge, mix up the design plans and hand them out again, making sure no one has his or her own plan. Ask students to find the model that matches the plan they were given.

Telephone Talk

Focus: Geometry and Language

INVESTIGATION 6

This problem is a variation on the classic game in which a student builds a design behind a barrier, and a partner tries to build the same design based on a description alone. This is always an interesting and challenging process, even for adults, because you find out how precise you need to be to get someone to replicate a design exactly. There are usually more attributes (size, shape, color, position) than you might think of at first.

Mathematical Ideas

- Describing line segments
- Describing geometric shapes
- Measuring length
- Measuring angles
- Describing location
- Analyzing patterns

Tools and Equipment

- ◆ blank paper
- ◆ lined writing paper
- ◆ pencils
- ◆ rulers
- ◆ compasses, protractors

Grouping Arrangement

Pairs

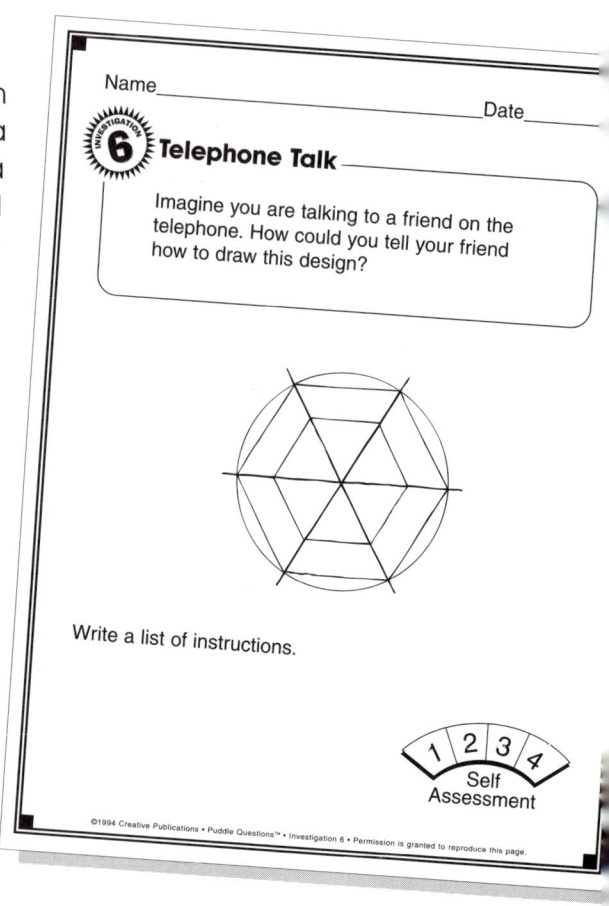

Name_____ Date_____

Telephone Talk

Imagine you are talking to a friend on the telephone. How could you tell your friend how to draw this design?

Write a list of instructions.

1 2 3 4
Self Assessment

©1994 Creative Publications • Puddle Questions™ • Investigation 6 • Permission is granted to reproduce this page.

Blackline Master in English, page
and in Spanish, page

Geometry and Language

Presenting the Problem

1. Pass out copies of the investigation and explain to the students that their task is to write a series of directions for drawing the design pictured. The directions should be so clear that a friend could draw the design exactly from hearing their description over the telephone. In order to write directions that will work, they should pay attention to the shapes in the design, the sizes of the shapes, lines, and angles that make up the design, and their locations in the design.

2. Students should imagine that the friend on the telephone would have blank paper, a pencil, and other supplies such as rulers, graph paper, or dot paper.

3. Students will work in pairs on this activity. Although both partners will know what the design looks like, they can take turns trying to draw the figure exactly the way their partner describes it. This might help the students revise their directions.

Assessment Criteria

It is important to let students know what they will be evaluated on in this assessment. Tell them that you are interested in finding out

✔ **whether a reader could draw the design from their description alone**

✔ **how precisely they describe**

- the lengths of the line segments in the design
- the sizes of the shapes, lines, and angles that make up the design
- the locations of shapes and lines in the design
- any patterns they notice in the design

Prompts for Getting Started

You may want to ask questions like these to help students get started:

- **How will you tell your friend exactly what to draw and where to draw it?**
- **How will your friend know what size to make each part of the design?**
- **What tools should your friend use to draw the design?**

Telephone Talk

Assessing the Work

In this assessment, we take a look at how well students are able to describe a design. It is interesting how many different ways they find to describe how to draw the design—from the inside out or from the outside in, for example. Some students may choose to use specific tools such as protractors and compasses; others may be less precise in their measurements. As you score the work, you might want to make a sketch following the student's directions.

Questions to ask yourself while scoring a response:

- If you drew the design the student describes, how closely would it match the design in the problem?
- What shapes does the student identify in the design?
- Does the student describe angle measurements and the length of the line segments?
- How well does the student describe the position of lines and shapes?
- Does the student recognize a pattern in the design?

SCORING RUBRIC

Low Response
A reader would have little success reproducing the design from the description.

Medium Response
A reader might have some success reproducing parts of the design from the description, which is more ambiguous than precise. The student may depend more on words like *short* and *across* to convey meaning and less on precise measurements.

High Response
A reader would be successful in reproducing a design similar to the one in the problem, though some aspects of the design are not described precisely. The student may have paid attention to some elements of the design but not others.

Exceptional Response
A reader would be successful in reproducing the design in the problem, possibly with a few modifications. Though it may not be perfect, on the whole the description is precise, unambiguous, and well presented. Care has been taken to make a detailed and complete description.

Geometry and Language

Supplies: Paper, pencil, ruler, and a protracter

Instructions

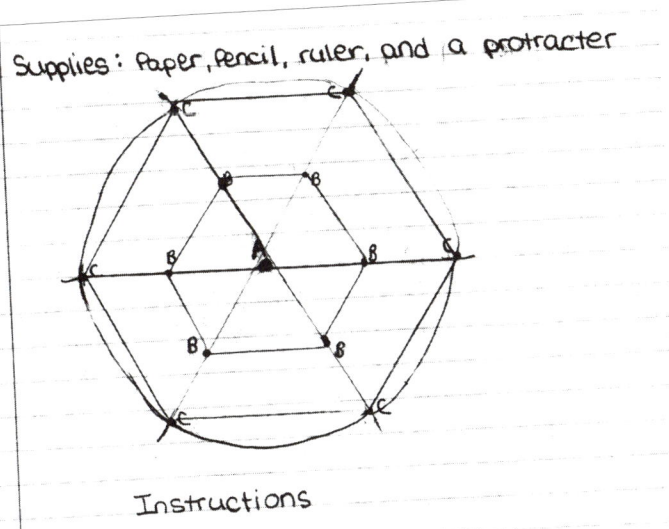

1. Draw a 4 inch line in the center of the page.
2. Find the center of the line and mark it with a dot. Name that dot A.
3. Draw 2 other 4 inch lines through point A at 60 degree angle.
4. Now on all the lines mark point B 1 inch from point A.
5. Now connect all the point 'B' with a ruler
6. You will now have a regular hexegon of all equal sides

7. Now on all the lines mark point C 2 inches from point A.
8. Connect point C with a ruler.
9. You have another hexegon of all equal sides.
10. Finally with point A as a center connect all of point C with a circle. You may want to erase the letters.

Exceptional Response

In this elegant response, the student gives a clear and concise set of directions. He uses lettered reference points that help make the description easy to follow.

Telephone Talk

Exceptional Response

This student takes a more conversational approach by describing the figure to a specific friend. Her description includes the use of both a compass and a protractor in drawing the circle and determining the angles of the line segments. Measurements are precise and even include the size of the paper.

> Investigation 6
> First of all Jeff, the materials you need are a ruler, a compass, a protractor a pencil, and a piece of paper at least 10 cm. across and 9½ cm. up and down.
> Ok now we are ready to begin. Listen closely to my instructions. First you draw a circle with the compass 4½ cm. radious around. From there you make a strait line across the center of the circle 10½ cm. long. Then you draw a 10½ cm. line at 60° angle with your protractor. Then draw another diagonal line at a 60° angle (on the other side of the line) from the first line with your protractor. The resulting diagram would look like a pie cut into 6 equal pieces. Make the lines you have placed onto the circle dark. Conect each point (around the circle) that the line crosses the circle. Making it into a hexigon. Make a point 1½ cm. inside the circle (from the points of the hexigon). Conect those points with strait lines. Forming a smaller hexigon in the larger hexigon.
> You then will be finished. It will look like a circle with an X on it with a line through the X. Then two hexigons in the circle. A big hexigon and a small hexigon.

High Response

Although this student took care to create step-by-step instructions, her overall strategy is not as effective as that in the exceptional responses, and the language is less precise. In one instance, she gives an angle measurement incorrectly (the 30-degree angle should be a 120-degree angle, twice a 60-degree angle). However, with a few clarifications, a closely matching figure could be drawn from this description.

> 1) Draw an X with lines that are 10.5 cm long. Make it have 60° angles at the top and bottom of the X. Make it have 30° angles on the sides of the X.
> 2) Draw a horisontal line where the two lines of the X meet. Make it 10.5 cm long.
> 3) Measure 7.8 cm on the horizontal line.
> 4) Make a line from that spot to the line in the X that is pointing Northeast. Make it 2.5 cm long. Measure the line pointing Northeast and make the line that is 2.5 cm long connect to it where it is 7.8 cm long.
> 5) From the end of that line make another 2.5 cm line the same way. Make 4 more like this. When you finash you should have a hexigon.
> 6) Draw another hexigon the same way except make it 1.5 cm larger.
> 7) Then take a protractor and put the pointed part where the 2 lines from the X and the horisontal line meet and draw a 4 cm circle.
> 8) When you are finash you should have a circle with 2 hexigons inside. You should also have an X with a horisontal line through it.

Geometry and Language

Medium Response

This set of instructions includes measurement information, but locational hints are lacking or ambiguous. The use of the star symbol on a telephone to describe the intersecting line segments in the figure is an effective, if not mathematical, strategy.

> Investigation #6
>
> 1st draw a hexagon exactly 4cm long for each of every six lines. Then make another hexagon smaller witch is exactly 1.6 cm long for each line. Look at your telephone on the bottom left hand corner, there is a ster, overlap the drawing with that ster making the lines exactly 10½ cm long. Get a compass and measure 1½ from the middle of the desighn around the bigger hexagon. You should come up with something like this... (but much bigger)

Low Response

Neither size nor location of the shapes is described precisely in this response. For example, the student says that the "X isn't very big," and the "hectagon is between the lines." He wants to go show his friend the figure, an acknowledgment that his list of five steps is probably not adequate.

> Hi frend guess what I found, I found a circle. I will explain it to you.
>
> Step 1. first the circle starts with an x, this x isn't very big.
>
> 2. it has a line through the x in the middle of the x. I will show it to you after this call. ✶ Its' like a pound sign
>
> 3. then it has a little hectogon in between the lines ✶
>
> 4. next it has another hectogon on the lines. ⬡
>
> 5. last it has a round circle over the hectogons and lines. ⊛
>
> I'll be over ther in about 10 minutes to show it to you.

Telephone Talk

Extending the Learning

1. After the students have completed their investigation, have them take their directions home and read them to a parent or other adult. **Have one of your parents draw the design the way you described it. Don't show the design until the drawing is finished. What your parent draws might help you to think of a better way to describe the design.** You might want to have students bring to school the designs from home so they can compare them to the original design.

2. Once the students have had a chance to try out their design description on an unknowing adult, gather the class together to discuss what happened. **What did you learn about giving the directions? How could you change your description so that it works better?**

3. As a whole group, devise a list of directions for drawing the design, incorporating everyone's additions and changes. As you write the directions, draw the figure on the chalkboard just as the student describes it, showing how someone might misinterpret it.

~~ Talking It Over ~~

What did you learn when you tried your directions out on an adult? Did you find any parts of the directions that should be changed?

~ My directions weren't specific enough. I started with the X, but I didn't say how big to make it. If I'd used more measurements, I think it would have gone better.

~ My mom tried making the design, and she came up with a square inside the circle instead of a hexagon. Next time, I'd say more about whether to draw horizontal, vertical, or diagonal lines.

~ Yeah. I realized I'd called the hexagon an octagon. That really messed things up.

~ My directions were really good at first. Then I think I got lazy and didn't use such specific instructions.

What did you learn when you shared your directions with a partner? How were your directions different?

~ I started with the X in the middle. My partner started with the circle.

~ To draw the lines, my partner said to put dots around the circle, then connect the dots. That worked really well.

• • •

Geometry and Language

Follow-Up Activity

1. Have the students make their own designs, without letting anyone else see what they are doing. They may use rulers, compasses, or protractors to construct the design, but they should make it simple enough that they will be able to describe it to a partner.

2. When students have made their design, have them pair up with a partner and give each other oral directions to reproduce the design. The partner should make the design exactly as directed, working in pencil. The direction-giver can look at the partner's drawing, but not vice versa. And the direction-giver must use words only—no gestures, pointing, or drawing in the air allowed.

3. As the students work, the direction-giver may notice that the partner's drawing does not exactly match the original. This may be because the partner is not following the directions exactly or because the directions are not clear enough. The direction-giver can repeat the directions or change them if necessary until the partner is able to draw the design correctly.

4. Comparing the drawn design with the original is always revealing.

Is It Fair?

Focus: Probability

In this investigation, students wrestle with the concepts of chance events and possible outcomes in a game setting. They are to explain whether a game is fair—it isn't—and how they can change it to make it fair. As the students play the game, they may discover that the game results are very close. Only by a careful analysis of the possible outcomes can they determine with certainty which player has a better chance of winning.

Mathematical Ideas

- Understanding chance events
- Interpreting data
- Analyzing possible outcomes
- Understanding relationships among operations

Tools and Equipment

◆ blank paper
◆ lined writing paper
◆ pencils and paper clips for spinners

Grouping Arrangement

Pairs

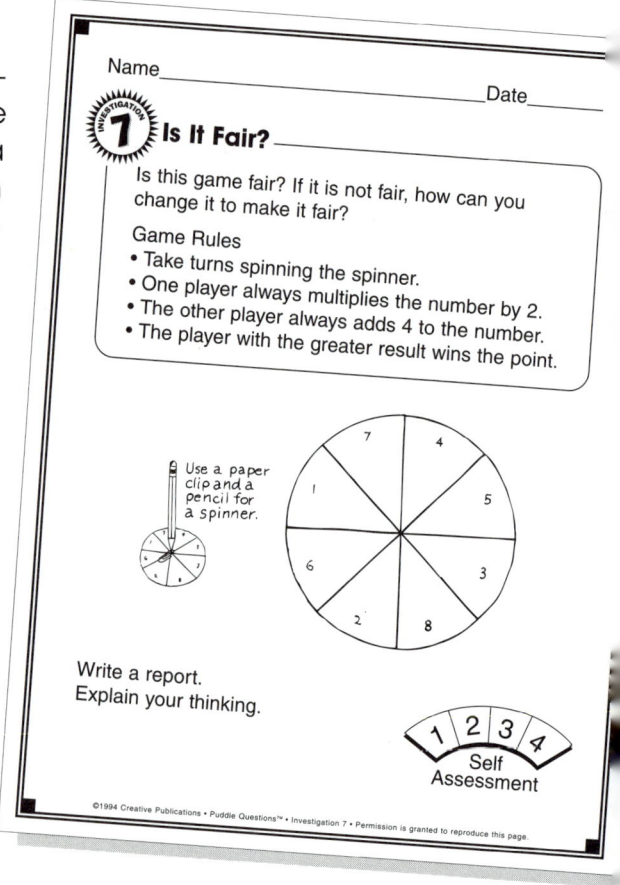

Name_____ Date_____

Is It Fair?

Is this game fair? If it is not fair, how can you change it to make it fair?

Game Rules
- Take turns spinning the spinner.
- One player always multiplies the number by 2.
- The other player always adds 4 to the number.
- The player with the greater result wins the point.

Use a paper clip and a pencil for a spinner.

Write a report. Explain your thinking.

1 2 3 4
Self Assessment

©1994 Creative Publications • Puddle Questions™ • Investigation 7 • Permission is granted to reproduce this page.

Blackline Master in English, page 85 and in Spanish, page 96

Probability

Presenting the Problem

1. Demonstrate how to use a pencil and a paper clip for a spinner and how to play the game described in the investigation. You might want to do this on the overhead projector, using a transparency of the investigation. The game is for two players. For each round of play, they spin the spinner (they can alternate turns spinning). One player will multiply the number that comes up by 2. The other player will add 4 to the number. The player with the higher result wins a point for that round.

2. Make sure everyone understands that "fair" means that each player in the game has the same chance of winning.

3. Let the pairs of students play the game. They can play as long as necessary in order to understand the game and its outcomes.

4. After the game, each student should write a report explaining whether the game is fair or unfair and why. **Explain your thinking fully. You may want to use sketches, charts, tables, or lists to make your report clearer.** Students who decide that the game isn't fair should explain how they would change it to make it fair.

Assessment Criteria

It is important to let the students know what they will be evaluated on in this assessment. Tell them that you are interested in finding out

- ✔ how they decide if the game is fair
- ✔ if they decide the game is unfair, whether they are able to create a game that is fair
- ✔ how clearly they present their thinking

Prompts for Getting Started

If students have difficulty getting started on the problem, you can ask questions like these:

- **Which player is more likely to win? Why?**
- **How can you show what might happen in a game?**
- **What makes the game fair or unfair?**
- **If you think the game is unfair, what are some ways to change it to make it fair?**

Is It Fair?

Assessing the Work

In this assessment, we take a look at how students analyze game outcomes, reason logically, and support their conclusions. Students' display of data, explanation of the game's fairness, and suggestions for making a fair game will reveal much about their understanding of probability. You may find some students still struggling with the meaning of fairness in games. Some may hastily decide that the game is fair if they have a tied score after a few rounds of play. Some will see that the game is unfair but be unable to explain the reason, while others will analyze the outcomes and know exactly why the game is unfair.

Questions to ask yourself while scoring a response:

- Does the student think the game is fair? What reasons are given? Does the student compare possible outcomes for the two game players?
- Does the student change the game to make it fair? In what way?
- How clearly is the student's reasoning explained?

SCORING RUBRIC

Low Response
The response shows little or no understanding of the concept of chance or fairness. There may be no comparison of possible outcomes or attempt to change the game.

Medium Response
The response shows partial understanding of the concept of chance or fairness. The reasoning may be based on the results from playing the game or on a comparison of the possible outcomes, but the reasoning may be flawed. The student may suggest a way to change the game, although the change may not result in a fair game.

High Response
The response shows good understanding of the concept of chance or fairness. The reasoning is based on a correct analysis of the possible outcomes in the game. The student has at least partial success changing the game to make it fair.

Exceptional Response
The work has the qualities of a high response, and in addition it shows exceptional mathematical thinking for the grade level.

Probability

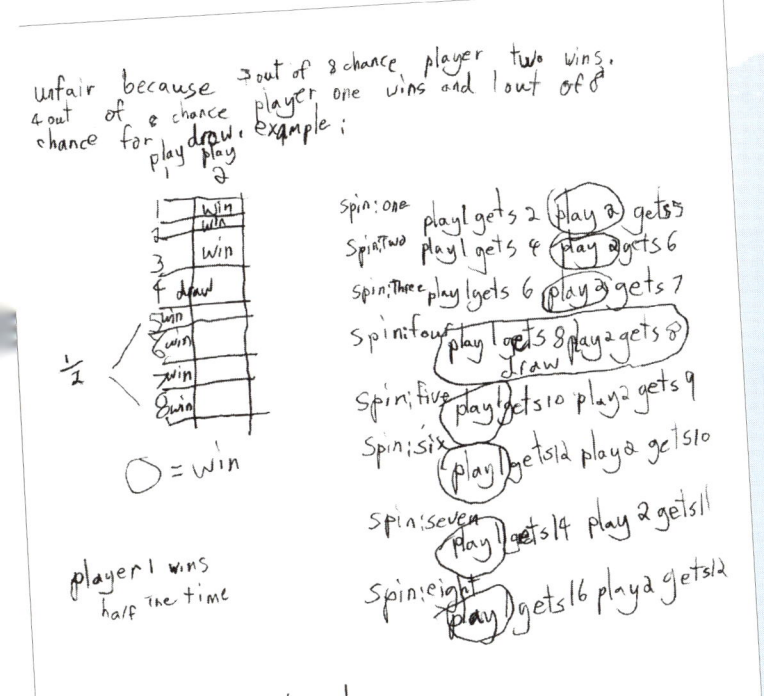

[Handwritten student work:]

unfair because 3 out of 8 chance player two wins.
4 out of 8 chance player one wins and 1 out of 8
chance for draw. example:
play play
1   2

1	win	
2	win	
3		win
4		draw
5	win	
6	win	
7		win
8	win	

½

○ = win

player 1 wins half the time

spin: one   play 1 gets 2   (play 2) gets 3
spin: two   play 1 gets 4   (play 2) gets 6
spin: three play 1 gets 6   (play 2) gets 7
spin: four   (play 1 gets 8 play 2 gets 8)
                          draw
spin: five   (play 1 gets 10) play 2 gets 9
spin: six   (play 1 gets 12) play 2 gets 10
spin: seven (play 1 gets 14) play 2 gets 11
spin: eight (play 1 gets 16) play 2 gets 12

play = player

To make fair: make player 2 get a point when theres a tie

Exceptional Response

This student clearly described his findings, and in addition accurately described the mathematical probability of each player's winning. He concluded that Player 1 would win half the time. Then he found an effective way to make the game fair—give the disadvantaged player a point when there would have been a tie.

High Response

This student uses a sketch of the spinner to give a clear picture of the probability of each player's winning. He then changes the number that results in a tie to a number that gives a win to Player C, making a fair game.

[Handwritten student work:]

B   C
|||| |||| |||| ||||
|||| ||||

This game is obviously not fair, even if you don't play, you can calculate it out, here's the explanation in the picture B 1st player, C 2nd player the X is both get a point, as you can see, Bs is more than Cs, so the posibility is bigger, all you got to do is change the 4 to 2, and it fair.

[spinner diagram with sections labeled: 7/4, 1 C, B 3 X, B 5, 6 B, C, B, C 3, 2, 8]

63

Is It Fair?

High Response

Redrawing the spinner with seven instead of eight sectors is a creative way to make the game fair. Having noted that × 2 has one more way of winning than + 4, this student simply removes one of the ways that × 2 can win.

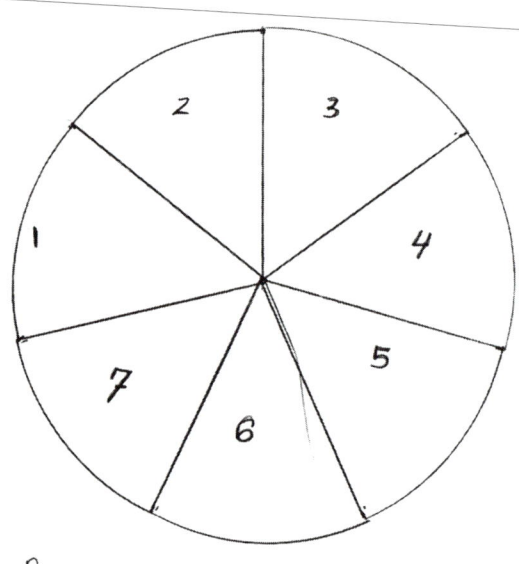

> No no es justo porque la persona con x2 tiene una oportunidad mas de ganar, si hacemos solo 7 pedasos Entonces funcionaria asi...
>
> 1 = +4 gana
> 2 = +4 gana
> 3 = +4 gana
> 4 = +4 & x2 gana ¡EMPATE!
> 5 = x2 gana
> 6 = x2 gana
> 7 = x2 gana
>
> al final acaba siendo justo!!!

English translation, page 76

> Yes, this game is Fair, because the numbers 4 and below count for the adding guy and the numbers 5 and up count for the multiply guy.

Medium Response

This student is in error about one outcome of the game. She has not seen that spinning a 4 results in a tie. If it were true that "the numbers 4 and below count for the adding guy," while "5 and up count for the multiply guy," the game would be fair, as she states.

Probability

Is this Game Fair?

Even though their should always Be a winner in a game, It should always be fair. All games have rules and this one as well as others have rules but it isn't fair.

It isn't fair because if one player multiplies by two during the whole game and the other player adds 4 for the whole game it wouldn't be fair. My partner and I played this game twice and the person who multyplied by 2 always wins.

I think if one time one player does multiplaction and then on his next turn does adding and so on it would be fair.

Medium Response

An analysis of the possible outcomes is not given in this response. The reasoning is based on experiential evidence, the fact that the player who multiplied by 2 won in two games.

I think this game is fair. Sure I think it needs some little changes but first I will tell you why I think it is fair. Nobody has an advantage. For example 2 times is 2 but 4 plus 1 is 5. But 2 times 4 is 8 but 4 plus 2 is six. Sometimes you win sometimes you don't. But about those changes, I think you should have the same number, not 2 numbers. Other than that the game was fun.

Low Response

The conclusion here is that "sometimes you win, sometimes you don't." The student's recommended change for the game, having the numbers be the same ($\times 2$ and $+ 2$, for example) would only increase the advantage of the player who multiplies.

Is It Fair?

Extending the Learning

1. Gather the students together to share their work and the thinking they did in this investigation. You may find yourself sitting back as the students discuss the questions: **What makes a game fair? Was this game fair or unfair?**

2. Have the partners tell what scores they got when playing the game. Record the scores for × 2 and + 4 on the chalkboard. **What do you notice about the scores?** Scores for the games may be very close, even though the probability of × 2 winning is greater.

3. Ask, **What were the possible outcomes of the spins?** Make a list of suggestions on the chalkboard. Students may suggest ways of organizing the data so that you have all the possible spins listed.

4. Finally, let students share different ways they found to change the game to make it fair. If possible, show each proposed change to the spinner on the overhead projector. Don't confine the discussion to ideas the students listed in their reports. Once they start brainstorming, they may come up with entirely new ideas.

~~ Talking It Over ~~

Do you think the game was fair or unfair? Why?

~ It was fair. We both won once.

~ It wasn't fair, I don't think. The player who multiplied by 2 won more often. Maybe you both won because the player who added 4 just got lucky.

~ Someone can always win a game, but the game should still be fair.

~ Yeah, it depends on what you spin. Sometimes one player wins, sometimes the other. It goes back and forth.

~ But this game *isn't* fair because if you play a lot, the player who multiplies has a better chance to win. So you can win most of the time just by choosing that side.

~ You can figure that out without ever playing the game. We made a chart that shows all the different numbers you could get, so we could see the game wasn't fair from the beginning.

• • •

Probability

Follow-Up Activity

1. Have students work in pairs or small groups to design a new spinner and a new set of rules for a fair game. Each player must perform a different operation, as was done in the investigation. Students might have players perform more than one operation, but caution them to make game rules that can be explained clearly.

2. Students should test their new games and make any revisions and refinements they want. Then they should make their spinners and a list of the rules of their game. When they are happy with the result, they can trade with another pair and have fun playing each other's games.

3. If students have any disagreements about whether games are fair, they should get together with the pair that made the game and discuss their ideas until they reach a consensus.

Lemonade Stand

Focus: Situational Problem Solving

Unlike the word problems in many textbooks, situational problems are complex, require some analysis, and do not have only one correct answer. They reflect the kind of real-world problem solving people use day to day. In this problem, students are asked to plan a lemonade stand, complete with details of cost and profit. Students can approach this problem in more or less detail, depending on whether they calculate, for example, the number of ounces per serving of lemonade or the cost to make one serving.

Mathematical Ideas

- Analyzing a situation
- Working with money
- Calculating profit

Tools and Equipment

- blank paper
- lined writing paper
- pencils, markers, crayons
- calculators
- samples of cups and lemonade with prices (optional)

Grouping Arrangement

Pairs or small groups

Name_____ Date_____

Lemonade Stand

Make a plan for selling lemonade.

Write a report about your plan. Predict how many servings you will sell. How much will you spend on supplies? How much will you charge? How much will you make?

Explain your thinking. Show all your calculations.

Self Assessment: 1 2 3 4

©1994 Creative Publications • Puddle Questions™ • Investigation 8 • Permission is granted to reproduce this page.

Blackline Master in English, page
and in Spanish, page

Presenting the Problem

1. Distribute the problem sheets. Ask students to think about the investigation for a moment. Then call on individuals to describe the task in their own words.

2. Make sure everyone understands that there is no one correct way to carry out this investigation. There are many different possibilities for planning the lemonade stand. For example, students can work with exact amounts of money or with rounded-off close estimates.

3. You might want to have available some different ways students can make lemonade: frozen lemonade in different size cans, a container of powdered lemonade, boxes of paper cups in different sizes, measuring cups, and pitchers in different sizes. Make sure there is a price on each item. Alternatively, provide the following information:

> can of frozen lemonade (makes 2 quarts)—$0.99
> box of 100 paper cups—$2.45

4. Although students will work in pairs or small groups, each student should prepare a report. Students will need calculators to do their figuring.

Assessment Criteria

It is important to let students know what they will be evaluated on in this assessment. Tell them that you are interested in finding out

- ✔ how clearly they show their plan for the cost and profit of the stand
- ✔ whether their plan is reasonable
- ✔ how clearly they show their math thinking

and to a lesser degree

- ✔ whether their calculations are correct

Prompts for Getting Started

As students work, you might want to ask individuals questions such as these:

- How can you show clearly what your costs will be to set up the stand?
- How will you figure out how much you are likely to make?
- Does your report make your mathematical thinking clear?

Lemonade Stand

Assessing the Work

In this assessment, we are interested in finding out how students approach a multifaceted problem, break it down, carry out the individual tasks, and organize information. Their reports will give you valuable insights into their ability to use mathematical calculations for a purpose.

Questions to ask yourself while scoring a response:

- How well does the response show the information asked for: the costs and the earnings for the lemonade stand?
- Is the student's proposed plan reasonable?
- Is the mathematical reasoning made clear in the report?
- Are the calculations correct?

SCORING RUBRIC

Low Response
Important information is missing from the response, so that it is unclear what the costs or profit would be. Or, figures given are completely unreasonable.

Medium Response
The basic information is presented, but the response may not make clear exactly what the total costs or the profit will be. There may be errors in the calculations.

High Response
The financial plan for the lemonade stand is complete. Costs, income, and profit are given. The reasoning is made clear by the presentation.

Exceptional Response
The response has all the qualities of a high response, and in addition reveals advanced mathematical thinking for the grade level, shown in the quality and/or organization of the work.

Situational Problem Solving

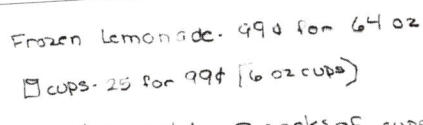

[Student work sample 1:]

Frozen lemonade - 99¢ for 64 oz
cups - 25 for 99¢ (6 oz cups)

First I would go and buy 2 packs of cups for about $2.00. Then I would buy about 5 Frozen Lemonades for about $5.00. Next I would set up my lemonade stand with a few cardboard boxes. Finally I would sell the lemonade for about 20¢ a cup

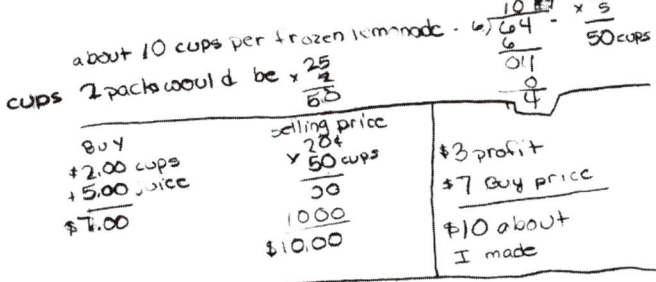

about 10 cups per frozen lemonade -
$$6\overline{)64} = 10\;R4 \qquad \times \begin{array}{c}10\\5\\\hline 50\text{ cups}\end{array}$$

cups 2 packs would be $\times \begin{array}{c}25\\2\\\hline 50\end{array}$

Buy	Selling price	
$2.00 cups	20¢ × 50 cups	$3 profit
$5.00 juice	$10.00	$7 Buy price
$7.00		$10 about I made

I would probably have my lemonade stand on the day there are lots of garage sales by my house!

Exceptional Response

All important details have been considered in this response. The student has calculated how many servings each container of frozen lemonade will yield, a practical way to approach the problem. Although the prices are rounded off, they are realistic; the student is likely to make the estimated profit.

Exceptional Response

The entire plan for a lemonade stand is communicated through this illustration. The grocery receipt shows what has been purchased and the total cost. The signs on the stand tell how much lemonade per glass and the cost, and the speech bubbles tell the rest.

71

Lemonade Stand

High Response

This student has analyzed the problem thoroughly. Because of her organization, it takes some work to follow the reasoning. She figures that she can make 32 one-quarter-cup servings from a 2-quart can of frozen lemonade. For cups and frozen lemonade, she will spend $5.42. With a possible $9.60 in earnings, the profit will be $4.18.

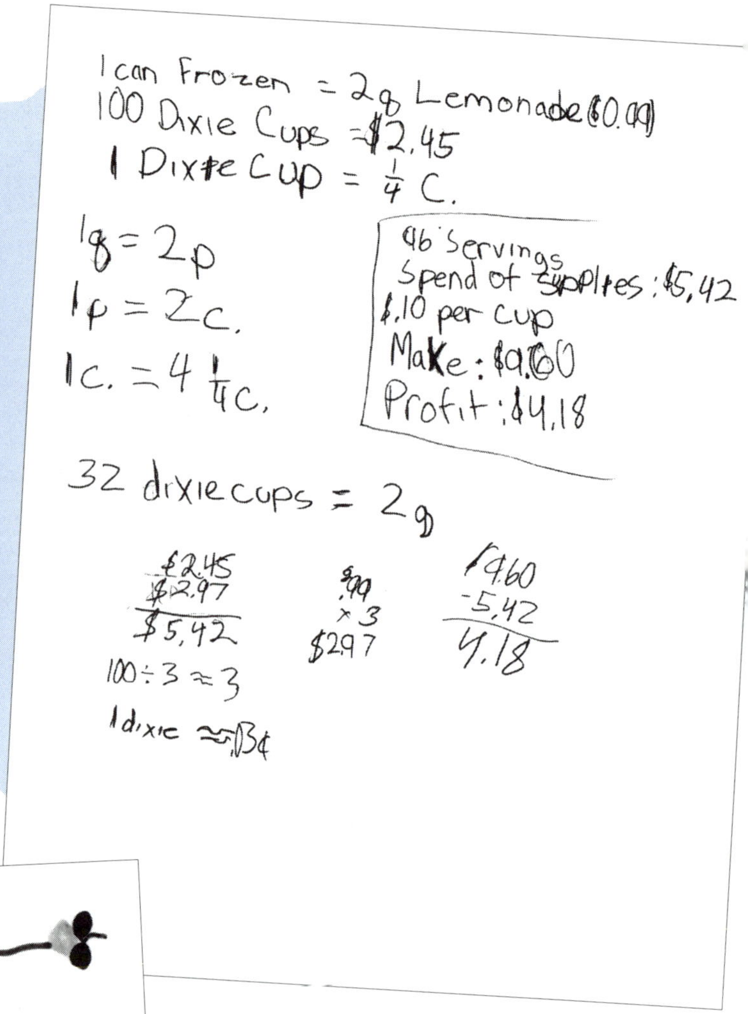

Student work:

1 can Frozen = 2q Lemonade ($0.99)
100 Dixie Cups = $2.45
1 Dixie Cup = $\frac{1}{4}$ C.

1q = 2p
1p = 2c.
1c. = 4 $\frac{1}{4}$c.

96 Servings
Spend of supplies: $5.42
$.10 per cup
Make: $9.60
Profit: $4.18

32 dixie cups = 2q

$2.45
$2.97

$5.42

$.99
× 3

$2.97

$9.60
-5.42

4.18

100 ÷ 3 ≈ 3
1 dixie ≈ 3¢

Medium Response

There are some clever marketing strategies in this response, but the cost list doesn't reflect the free treats. It looks as if the student finds herself in the hole because she didn't figure her income over the seven days she will be in business.

Student work:

If I would to sell lemonade it would cost, $3.00 for water, $2.00 for frozen lemonade, and $4.00 for cups. So I would spend a total of $9.00. I would have two size cups. One would be large and one would be small. Large for 75¢ and small for 25¢. If someone would by 5 cups of any size they would get a free cookie. And for the 10th customer would get a free cookie and lemonade. If 10 people bought any size and I sold lemonade for 7 days I would get about 2.50-7.50. So I would still be $4.00 short. So I would have a sale.

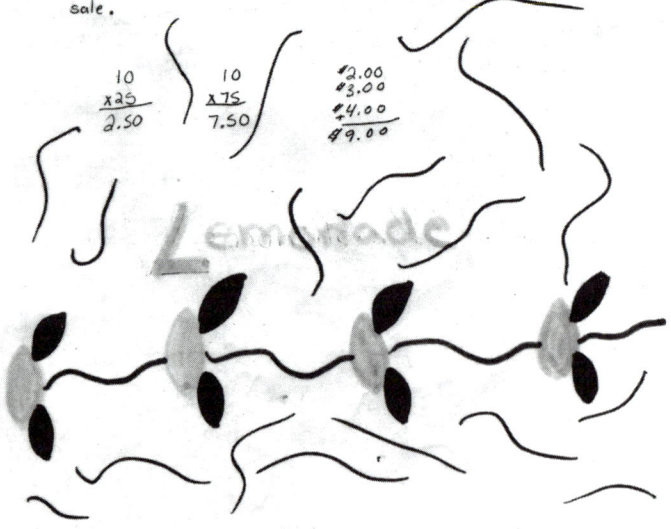

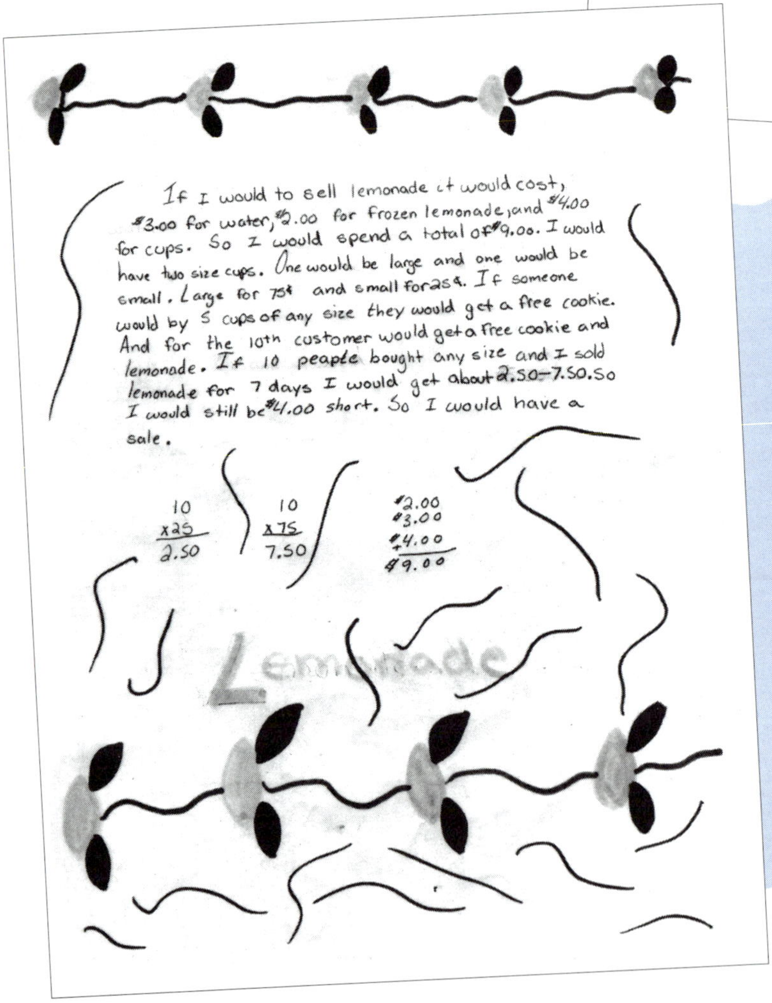

10
×25

2.50

10
×75

7.50

$2.00
$3.00
$4.00

$9.00

72

Situational Problem Solving

> I plan on spending 10.00 dollars on the supplies for the Lemonade stand. the cups of lemonade will cost 25¢. I am planing on selling 5 cups per half hour. In three hours I would have $12.00.
>
> LEMONADE
>
> If I sale it for 5 hours I will make $20.00. which is not bad for half a days work.

Medium Response

This student has calculated how much the lemonade stand will make if it stays in business for three hours or five hours. She tells how much she will spend on supplies, but does not itemize the supplies. She leaves the reader to figure the profit.

Low Response

The amounts given in this response are not related to each other to show how the fifteen-dollar figure is derived.

> Frozen Lemonad
> 25 cups for 99¢ [6 oz cups]
>
> I think I will serve about 19 cup in about 17 minuts 5$ worth of supplies and I will charge 10¢ a cup and 25¢ for Lemonad I think I will make about 15 dollors

Lemonade Stand

Extending the Learning

1. When students have finished their reports, hold a class discussion about their results. **What questions did you have to ask in order to figure out how much money your lemonade stand would make?** List the students' questions on the chalkboard.

2. When the class agrees that the list is complete, ask, **How do you go about answering these questions? Is it possible to answer all of them exactly?** Discuss with students how they can get the information they need for this kind of real-life situation.

~~ Talking It Over ~~

What are some of the things you had to know in order to make your plan for the lemonade stand?

~ You had to know how much lemonade costs. There's different ways to buy it—like frozen or powder.

~ Or if you have lemons, you don't have to pay for them. But you have to get some sugar and find out how to make lemonade that way.

~ You have to know how much you can sell. I didn't know how to guess that.

~ I made a lemonade stand last year, so I remembered how that worked and how much I could sell then.

How did you figure out how much to sell your lemonade for?

~ I just thought people wouldn't buy it if it was too much, so I said 25¢.

~ It's better if you figure out what you want to make. Like if you want to make $5.00 and you already spent $7.00, then you have to sell $12.00 worth of lemonade.

~ I figured out I had to spend 3¢ for each cup of lemonade, so if I sold a cup for 25¢, I knew how much I made each time.

• • •

Situational Problem Solving

Follow-Up Activity

1. Discuss with the students factors to consider in figuring the profit from a lemonade stand. They might suggest, for example, that weather, season, day of the week, number of hours open for business, or location would be factors.

2. Have the students explore at least one of the factors and how the expected profit would change depending on various conditions. For example, the students might explore how different types of weather would affect profits. They could make a spreadsheet for the factors, either by hand or with a spreadsheet program if one is available to your class. Students could work in pairs, small groups, or as a class.

3. Display the completed spreadsheets for everyone to see.

75

Bibliography

Ann Arbor Public Schools. *Alternative Assessment: Evaluating Student Performance in Elementary Mathematics.* Palo Alto, Calif.: Dale Seymour, 1993.

California Department of Education. *A Sampler of Mathematics Assessment.* Prepared by Tej Pandey. Sacramento, 1991.

California Mathematics Council and EQUALS. *Assessment Alternatives in Mathematics.* ed. Jean Kerr Stenmark. 1989.

Freedman, Robin Lee Harris. *Open-Ended Questioning: A Handbook for Educators.* Menlo Park, Calif.: Addison-Wesley, 1994.

Hart, Diane. *Authentic Assessment: A Handbook for Educators.* Menlo Park, Calif.: Addison-Wesley, 1994.

Mathematical Sciences Education Board, National Research Council. *Measuring Up: Prototypes for Mathematics Assessment.* Washington, D.C.: National Academy Press, 1993.

National Council of Teachers of Mathematics. *Assessment in the Mathematics Classroom: 1993 Yearbook.* eds. Norman L. Webb and Arthur F. Coxford. Reston, Va., 1993.

National Council of Teachers of Mathematics. *Mathematics Assessment: Alternative Approaches.* Reston, Va., 1992. Videotape and guidebook.

National Council of Teachers of Mathematics. *Mathematics Assessment: Myths, Models, Good Questions, and Practical Suggestions.* ed. Jean Kerr Stenmark. Reston, Va., 1991.

Translations of Student Work

Investigation #1, Page 15
1. meterstick—I would use a meterstick to measure across the puddle.
2. pencils—I would use pencils to measure the width of the puddle. 3. shoes—I would use shoes to measure across the puddle. 4. string—I would use a string to measure around the puddle. 5. paper clips—I would use paper clips to measure across the puddle. 6. ruler—I would use a ruler to see how deep a puddle is. 1. meterstick; 2. pencils; 3. shoes; 4. string; 5. paper clips; 6. ruler

Investigation #7, Page 64
No it is not fair because the person with ×2 has one more chance of winning. If you make only 7 pieces then it works out like this…1 = +4 wins; 2 = ÷4 wins; 3 = +4 wins; 4 = +4 and ×2 win—tie; 5 = ×2 wins; 6 = ×2 wins; 7 = ×2 wins. It ends up fair!

Name_____ Date_____

The Puddle Problem

How would you measure a puddle?

Record all the different ways you can think of. Make sketches to show your ways.

Self Assessment

Nombre_____ Fecha_____

Investigación 1: El problema del charco

¿Cómo medirías un charco?

Anota todas las maneras que se te ocurran. Luego haz dibujos que ilustren tus ideas.

| 1 | 2 | 3 | 4 |

Evaluación propia

Name_____ Date_____

Television Survey

Conduct a survey of your class about television.

Think of a question you would ask the students in your classroom. Conduct the survey. Write a report about your results. What did you learn from the survey?

Self Assessment: 1 2 3 4

©1994 Creative Publications • Puddle Questions™ • Investigation 2 • Permission is granted to reproduce this page.

Nombre_____ Fecha_____

INVESTIGACIÓN 2: Encuesta sobre la televisión

Lleva a cabo una encuesta en tu clase sobre la televisión.

Piensa en una pregunta que te gustaría hacerles a tus compañeros de clases. Lleva a cabo la encuesta. Luego, escribe un informe acerca de los resultados. ¿Qué aprendiste de la encuesta?

Evaluación propia: 1 2 3 4

Name_____ Date_____

Investigation 3: One to a Million

About how long would it take you to count to one million?

Record your estimate and explain your reasoning. Include all of your figuring.

Self Assessment: 1 2 3 4

©1994 Creative Publications • Puddle Questions™ • Investigation 3 • Permission is granted to reproduce this page.

Nombre_____ Fecha_____

Investigación 3: De uno a un millón

Aproximadamente, ¿cuánto tiempo te tomaría contar hasta un millón?

Anota tu estimación y explica tu razonamiento. Incluye todos tus cálculos.

Evaluación propia: 1 2 3 4

Name_____ Date_____

Investigation 4: Equations Galore!

What equations can you write for $\frac{1}{2}$?

Think of lots of different kinds of equations for $\frac{1}{2}$. Make lists of equations that are alike. What do you notice?

Self Assessment: 1 2 3 4

©1994 Creative Publications • Puddle Questions™ • Investigation 4 • Permission is granted to reproduce this page.

Nombre_____Fecha_____

4 ¡Ecuaciones por todos lados!

¿Qué ecuaciones puedes escribir para el $\frac{1}{2}$?

Piensa en distintos tipos de ecuaciones para el $\frac{1}{2}$. Haz listas de ecuaciones que se asemejen. ¿Qué observas cuando comparas las ecuaciones?

| 1 | 2 | 3 | 4 |

Evaluación propia

©1994 Creative Publications • Puddle Questions™ • Investigation 4 • Permission is granted to reproduce this page.

Name_____ Date_____

Investigation 5: Doghouse Designs

Design a doghouse made with plywood.

Plywood sheets are 4 ft by 8 ft.

Write a proposal explaining how you would build the doghouse. In the proposal, show a sketch of your design labeled to show its dimensions. Tell how much plywood you would need to build the doghouse. Why is your design a good one?

1 | 2 | 3 | 4
Self Assessment

©1994 Creative Publications • Puddle Questions™ • Investigation 5 • Permission is granted to reproduce this page.

Nombre_____ Fecha_____

INVESTIGACIÓN 5: Diseño de una casa para perros

Diseña una casa para perros con madera terciada o triplex.

Las láminas de madera terciada o triplex miden 4 pies por 8 pies.

Escribe una propuesta explicando como vas a construir la casa para perros. En la propuesta muestra un borrador de tu diseño, indicando sus dimensiones. Dí cuánta madera terciada o triplex necesitarás para construir la casa para perros. ¿Por qué crees que tu diseño es bueno?

Evaluación propia: 1 2 3 4

©1994 Creative Publications • Puddle Questions™ • Investigation 5 • Permission is granted to reproduce this page.

Name_____ Date_____

Telephone Talk

Imagine you are talking to a friend on the telephone. How could you tell your friend how to draw this design?

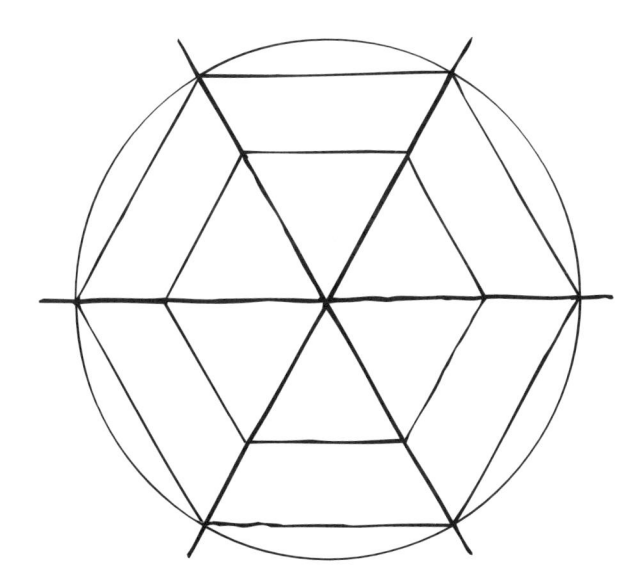

Write a list of instructions.

Self Assessment: 1 2 3 4

Nombre_____ Fecha_____

INVESTIGACIÓN 6 — Conversación telefónica

Imagínate que estás hablando por teléfono con un amigo. ¿Cómo le explicarías a tu amigo cómo dibujar este diseño?

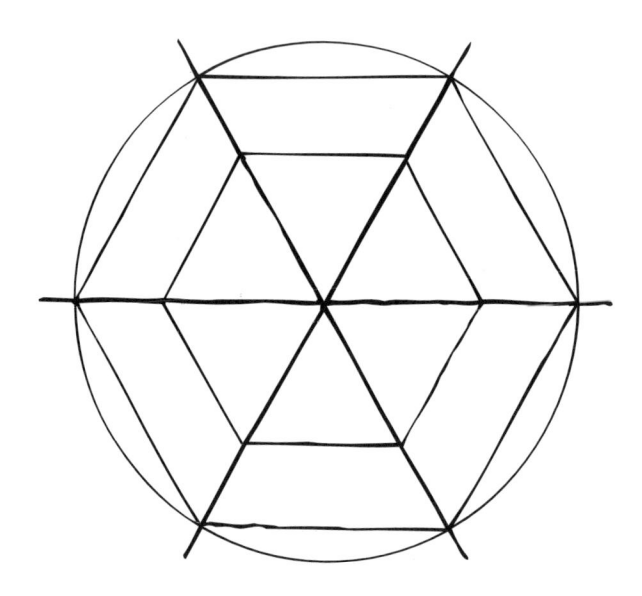

Haz una lista de instrucciones.

1 | 2 | 3 | 4
Evaluación propia

©1994 Creative Publications • Puddle Questions™ • Investigation 6 • Permission is granted to reproduce this page.

Name_____ Date_____

Investigation 7: Is It Fair?

Is this game fair? If it is not fair, how can you change it to make it fair?

Game Rules
- Take turns spinning the spinner.
- One player always multiplies the number by 2.
- The other player always adds 4 to the number.
- The player with the greater result wins the point.

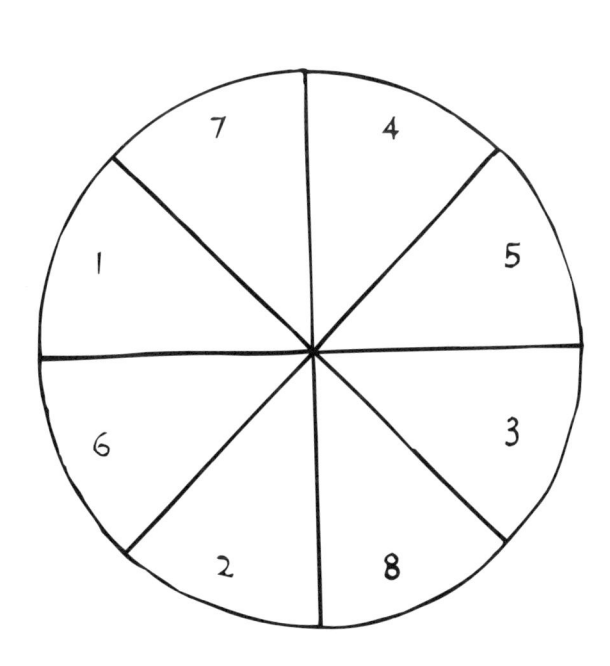

Use a paper clip and a pencil for a spinner.

Write a report.
Explain your thinking.

Self Assessment: 1 2 3 4

©1994 Creative Publications • Puddle Questions™ • Investigation 7 • Permission is granted to reproduce this page.

89

Nombre_____ Fecha_____

¿Es justo?

¿Te parece justo este juego? Si no lo es, ¿cómo lo cambiarías para que fuera justo?

Reglas del juego
- Túrnense para hacer girar la ruleta.
- Un jugador siempre multiplica el número por 2.
- Los demás siempre le suman 4 al número.
- Gana un punto el jugador con el mayor resultado.

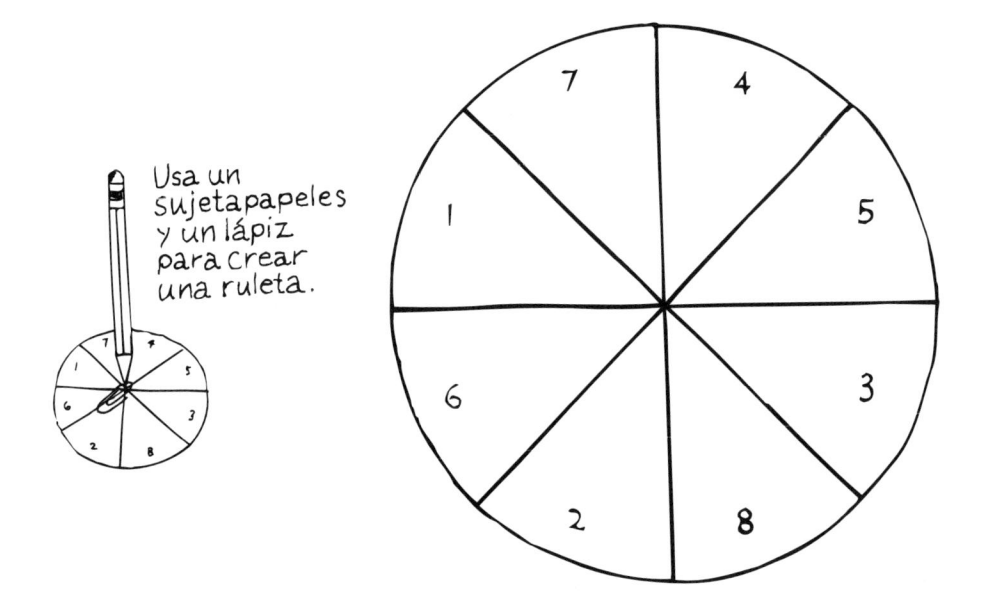

Usa un sujetapapeles y un lápiz para crear una ruleta.

Escribe un informe en el que expliques tu razonamiento.

Evaluación propia

©1994 Creative Publications • Puddle Questions™ • Investigation 7 • Permission is granted to reproduce this page.

Name_____ Date_____

Lemonade Stand

Make a plan for selling lemonade.

Write a report about your plan. Predict how many servings you will sell. How much will you spend on supplies? How much will you charge? How much will you make?

Explain your thinking. Show all your calculations.

1 2 3 4
Self Assessment

Nombre_____ Fecha_____

Investigación 8: La caseta de la limonada

Haz un plan para vender limonada.

Escribe un informe acerca de tu plan. Luego, haz una predicción del número de vasos de limonada que venderás. ¿Cuánto gastarás en provisiones? ¿Cuánto cobrarás por cada vaso de limonada? ¿Cuánto dinero ganarás?

Explica tu razonamiento y muestra todos tus cálculos.

Evaluación propia: 1 2 3 4

Writing Math Reports

Show what you know.

Do your best work.

Explain your thinking fully.

Make your thinking clear. A graph, table, chart, or picture might help.

Revise, if necessary.

Self-Assessment Key

How did I do?

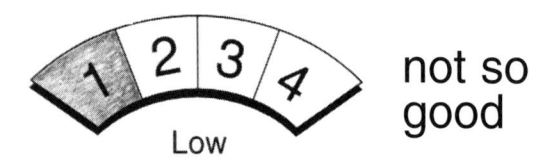

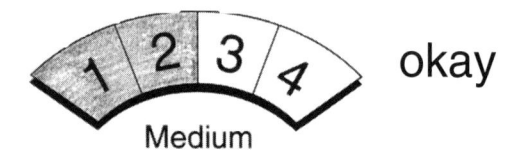

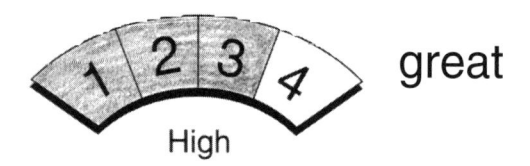

1 2 3 4 — Low	not so good
1 2 3 4 — Medium	okay
1 2 3 4 — High	great
1 2 3 4 — Exceptional	extra special

©1994 Creative Publications • Puddle Questions™ • Permission is granted to reproduce this page.

Observation Sheet   Investigation #_____

Name_____ Date_____

Notes:

Focus:
- ❏ Communication ❏ Attitude ❏ Collaboration
- ❏ Math Understanding ❏ Reasoning ❏ Use of Tools & Equipment

©1994 Creative Publications • Puddle Questions™ • Permission is granted to reproduce this page.

Observation Sheet   Investigation #_____

Name_____ Date_____

Notes:

Focus:
- ❏ Communication ❏ Attitude ❏ Collaboration
- ❏ Math Understanding ❏ Reasoning ❏ Use of Tools & Equipment

©1994 Creative Publications • Puddle Questions™ • Permission is granted to reproduce this page.

Record Sheet

Scoring Key
Low=1 High=3
Medium=2 Exceptional=4

Investigation

Student Names	Date	1	2	3	4	5	6	7	8
1.									
2.									
3.									
4.									
5.									
6.									
7.									
8.									
9.									
10.									
11.									
12.									
13.									
14.									
15.									
16.									
17.									
18.									
19.									
20.									
21.									
22.									
23.									
24.									
25.									
26.									
27.									
28.									
29.									
30.									
31.									
32.									
33.									
34.									
35.									

©1994 Creative Publications • Puddle Questions™ • Permission is granted to reproduce this page.